T0324492

ELEMENTARY
FUNCTIONAL
ANALYSIS

ELEMENTARY
FUNCTIONAL
ANALYSIS

Charles Swartz
New Mexico State University, USA

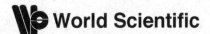

World Scientific

NEW JERSEY · LONDON · SINGAPORE · BEIJING · SHANGHAI · HONG KONG · TAIPEI · CHENNAI

Published by

World Scientific Publishing Co. Pte. Ltd.

5 Toh Tuck Link, Singapore 596224

USA office: 27 Warren Street, Suite 401-402, Hackensack, NJ 07601

UK office: 57 Shelton Street, Covent Garden, London WC2H 9HE

British Library Cataloguing-in-Publication Data
A catalogue record for this book is available from the British Library.

ELEMENTARY FUNCTIONAL ANALYSIS

Copyright © 2009 by World Scientific Publishing Co. Pte. Ltd.

All rights reserved. This book, or parts thereof, may not be reproduced in any form or by any means, electronic or mechanical, including photocopying, recording or any information storage and retrieval system now known or to be invented, without written permission from the Publisher.

For photocopying of material in this volume, please pay a copying fee through the Copyright Clearance Center, Inc., 222 Rosewood Drive, Danvers, MA 01923, USA. In this case permission to photocopy is not required from the publisher.

ISBN-13 978-981-4273-34-3
ISBN-10 981-4273-34-1

Printed in Singapore.

Preface

This text is based on a set of lecture notes for an introductory course in functional analysis which was offered during a summer session at New Mexico State University. Most of the students had only completed a first year course in real analysis and had no background in Lebesgue integration. Although this lack of knowledge of the Lebesgue integral limits the application of the results in abstract functional analysis to problems in concrete function spaces, particularly in partial differential equations, there are still sufficient applications presented in the text to topics in ordinary differential equations, integral equations, harmonic analysis and numerical integration to illustrate the many applications of functional analysis. Only a very basic knowledge of linear algebra and introductory real analysis including a knowledge of metric spaces at the level of Apostol ([Ap]), Bartle ([Ba]), De-Pree/Swartz ([DeS]) or Rudin ([Ru]) is necessary to follow the developments in the text; at one point in the chapter on spectral theory, Liouville's theorem from complex analysis is used. In the first chapter, we introduce and establish the basic properties of normed spaces and in Chapter 2, we study continuous linear operators between normed spaces and give applications to integral equations. In Chapters 3, 4 and 5, we consider quotient spaces, finite dimensional normed spaces and inner product (Hilbert) spaces. In Chapters 6-12, we establish the 3 basic principles of functional analysis — the Hahn-Banach Theorem, the Uniform Boundedness Principle and the Open Mapping/Closed Graph Theorems — and give applications to Banach limits, normed spaces, Fourier series, summability, weak and weak* convergence, projections and Schauder bases. In Chapter 13, we define the transpose (conjugate, adjoint) of a continuous linear operator between normed spaces and establish some of the relationships between the range and kernel of an operator and its transpose. In Chapter 14, we introduce the class of compact operators and in Chapter 15, we establish the Fredholm Alternative for compact operators. The remaining chapters of the text are devoted to studying spectral theory for continuous linear operators and giving applications to ordinary differential and integral equations. Chapters 16 and 17 give the definition of the spectrum of a continuous linear operator and examples of spectrums. Chapter 18 contains a description of the spectrum of a compact operator while Chapters 19 and 20 define symmetric operators and establish the spectral theorem for compact, symmetric operators and give applications to integral equations. Chapter 21 gives a spectral representation for operators which have compact inverses and an application of the representation to Sturm-Liouville ordinary differential equations. Chapters 22-25 establish the spectral theorem for bounded self adjoint operators on a Hilbert space and Chapter 26 uses the spectral theorem to define an operational calculus for bounded self adjoint operators; the operational calculus is then employed to establish the existence of the square root of a positive operator and the polar decomposition for a normal operator.

There are a number of excellent texts on functional analysis. There are two things that distinguish this text from other texts in functional analysis. The first is that there are minimal prerequisites, especially there is no requirement of the knowledge of the Lebesgue integral or general topology. There are a number of functional analysis texts which require similar minimal backgrounds such as, Pryce ([Pr]) or Schecter ([Sch]), but these texts do not contain many applications of the abstract results of functional analysis to concrete problems in function spaces whereas we present a large number of applications to ordinary differential and integral equations, Banach limits, Schauder bases, harmonic analysis, summability and numerical integration. On the other hand, there are a large number of functional analysis texts such as Dunford/Schwartz ([DuS]), Rudin ([Ru2]), Taylor/Lay ([TL]), and Yosida ([Y]) which contain many applications of the abstract results of functional analysis to function spaces but which have much more stringent background requirements.

Contents

Preface   v

1. Normed Linear and Banach Spaces   1

2. Linear Operators   9

3. Quotient Spaces   21

4. Finite Dimensional Normed Spaces   25

5. Inner Product and Hilbert Spaces   29

6. The Hahn-Banach Theorem   41

7. Applications of the Hahn-Banach Theorem to Normed Spaces   47

8. The Uniform Boundedness Principle   53

9. Weak Convergence   61

10. The Open Mapping and Closed Graph Theorems   69

11. Projections   73

12. Schauder Basis   77

13.   Transpose and Adjoints of Continuous Linear Operators   83

14.   Compact Operators   91

15.   The Fredholm Alternative   97

16.   The Spectrum of an Operator   103

17.   Subdivisions of the Spectrum   107

18.   The Spectrum of a Compact Operator   111

19.   Symmetric Linear Operators   115

20.   The Spectral Theorem for Compact Symmetric Operators   119

21.   Symmetric Operators with Compact Inverse   129

22.   Bounded Self Adjoint Operators   133

23.   Orthogonal Projections   141

24.   Sesquilinear Functionals   143

25.   The Spectral Theorem for Bounded Self Adjoint Operators   145

26.   An Operational Calculus   151

27.   The Spectral Theorem for Normal Operators   159

Appendix A   Functions of Bounded Variation   163

Appendix B   The Riemann-Stieltjes Integral   167

Appendix C   The Dual of $C[a,b]$ 169

Appendix D   The Baire Category Theorem   173

References   175

Index   179

Chapter 1

Normed Linear and Banach Spaces

In this chapter, we give definitions and examples of normed and Banach spaces.

Let $\Bbbk$ be either the field of real, $\mathbb{R}$, or complex, $\mathbb{C}$, numbers and let X be a vector space over $\Bbbk$.

Definition 1.1. A semi-norm on X is a function $\|\cdot\| : X \to [0, \infty)$ satisfying the triangle inequality $\|x + y\| \le \|x\| + \|y\|$ and $\|tx\| = |t|\,\|x\|$ for all $x, y \in X$ and $t \in \Bbbk$.

Note that we have $\|x\| = \|-x\|$ and if $x = 0$ in X, then $\|x\| = 0$ since $\|0x\| = 0$ by the second condition; if $\|x\| = 0$ iff $x = 0$, then $\|\cdot\|$ is called a *norm*. If $\|\cdot\|$ is a semi-norm on X, then $d(x, y) = \|x - y\|$ induces a semi-metric on X which is a metric iff $\|\cdot\|$ is a norm. Unless otherwise stated we always assume that X is equipped with this semi-metric (metric).

If $\|\cdot\|$ is a semi-norm (norm) on X, the pair $(X, \|\cdot\|)$ is called a semi-normed (normed) space; if the semi-norm (norm) is understood, we call X a semi-normed (normed) space.

We have the following topological properties of the semi-metric (metric).

Proposition 1.2. *Let X be a semi-normed space. Then*

(i) *the map $(t, x) \to tx$ from $\Bbbk \times X \to X$ is continuous,*

(ii) *the map $(x, y) \to x + y$ is continuous from $X \times X \to X$.*

Proof: (i): $\|tx - sy\| \le |t - s|\,\|y\| + |t|\,\|x - y\|$.

(ii): See Exercise 1 for the semi-norm on $X \times X$. Then we have

$$\|(x + y) - (x_0 - y_0)\| \le \|x - x_0\| + \|y - y_0\|$$

which gives the result.

Proposition 1.3. *Let X be a semi-normed space. If $x, y \in X$, then $\big| \|x\| - \|y\| \big| \le \|x - y\|$. Hence, the map $x \to \|x\|$ from X into $\mathbb{R}$ is continuous.*

Proof: $\|x\| \le \|x - y\| + \|y\|$ and $\|y\| \le \|y - x\| + \|x\|$ so $\pm(\|x\| - \|y\|) \le \|x - y\|$. The last statement follows immediately from the inequality.

We now give a number of examples of semi-normed and normed spaces.

Example 1.4. $\mathbb{R}^n$ with the euclidean norm, $\|x\| = \|(x_1, ..., x_n)\| = (\sum_{j=1}^{n} x_j^2)^{1/2}$, is a normed space.

Example 1.5. Let $S \neq \varnothing$ and let $B(S)$ be the space of all bounded functions $f : S \to \Bbbk$. Then $B(S)$ is a vector space where the operations of addition and scalar multiplication are defined pointwise and $\|f\|_\infty = \sup\{|f(t)| : t \in S\}$ defines a norm on $B(S)$ called the sup$-norm$. Note that convergence with respect to $\|\cdot\|_\infty$ is just uniform convergence on S. When $S = \mathbb{N}$, $B(S)$ is the space of all bounded sequences and is denoted by l^∞.

If it is necessary to distinguish between real and complex scalars, we use the notation $B_\mathbb{C}(S)$ or $B_\mathbb{R}(S)$ or we warn the reader which scalar field is being used. Similar notation is used for the function spaces below.

Example 1.6. Let S be a compact metric (topological) space and let $C(S)$ be the space of all $\Bbbk$ valued continuous functions on S. Then $C(S)$ is a subspace of $B(S)$ and is a normed space under the sup-norm.

Example 1.7. Let $C^1[a, b]$ be the subspace of $C[a, b]$ consisting of the functions with one continuous derivative. Besides the sup-norm on $C^1[a, b]$, we also have the norm $\|f\|_1 = \|f\|_\infty + \|f'\|_\infty$.

Example 1.8. Let s be the vector space of all $\Bbbk$ valued sequences (the vector operations of addition and scalar multiplication are defined coordinatewise). We consider several subspaces of s with the sup-norm. Of course, we have the space l^∞ of bounded sequences introduced in Example 5. Let m_0 be the space of all sequences which have finite range; let c be the space of all convergent sequences and let c_0 be the subspace of all sequences which converge to 0. Finally, let $c_{00}(c_c)$ be the space of all sequences which are eventually 0 (constant); that is, $t = (t_1, t_2, ...) = \{t_j\} \in c_{00}$ iff there exists n (depending on t) such that $t_j = 0$ (t_j is constant) for $j \ge n$. We assume that all of these spaces are equipped with the sup-norm.

Example 1.9. Let bs be the space of all bounded series; that is, all sequences $t = \{t_j\}$ such that $\|t\|_{bs} = \sup\{\left|\sum_{j=1}^{n} t_j\right| : n \in \mathbb{N}\} < \infty$. Let cs be the subspace of bs consisting of all convergent series.

Example 1.10. Let $BV[a,b]$ be the space of all functions $f : [a,b] \to \mathbb{R}$ with bounded variation (see Appendix A for the definition). Then $BV[a,b]$ is equipped with the variation norm $\|f\| = Var(f;[a,b]) + |f(a)|$ (see Appendix A).

Example 1.11. We next consider subspaces of l^∞ consisting of all p^{th} power convergent series; for $1 \le p < \infty$, define $l^p = \{\{t_j\} : \sum_{j=1}^{\infty} |t_j|^p < \infty\}$. We define a norm on l^p by $\|\{t_j\}\|_p = (\sum_{j=1}^{\infty} |t_j|^p)^{1/p}$. To show that $\|\cdot\|_p$ defines a norm on l^p, we establish two inequalities.

Theorem 1.12. *(Hölder's Inequality)* Let $1 < p < \infty$ and let q satisfy $\frac{1}{p} + \frac{1}{q} = 1$. Then $\sum_{j=1}^{n} |t_j s_j| \le (\sum_{j=1}^{n} |t_j|^p)^{1/p}(\sum_{j=1}^{n} |s_j|^q)^{1/q}$ for $t_j, s_j \in \mathbb{C}$.

Proof: First we show that

$$(\#) \quad a^{1/p}b^{1/q} \le \frac{a}{p} + \frac{b}{q}$$

for $a, b \ge 0$. Consider the function $f(t) = t^c - ct + c - 1$ for $0 < c < 1, t \ge 0$. Then $f'(t) = c(t^{c-1} - 1)$ so $f(1) = f'(1) = 0$ and $f'(t) > 0$ for $0 < t < 1$ and $f'(t) < 0$ for $t > 1$. Hence, $f(t) \le 0$ for $t \ge 0$. Now $(\#)$ is obvious for $b = 0$, so suppose that $b > 0$ and set $t = a/b$ and $c = 1/p$. Then $f(a/b) = (a/b)^{1/p} - \frac{1}{p}\frac{a}{b} + \frac{1}{p} - 1 \le 0$. Multiplying this inequality by b gives $a^{1/p}b^{1-1/p} = a^{1/p}b^{1/q} \le \frac{a}{p} + b(1 - \frac{1}{p}) = \frac{a}{p} + \frac{b}{q}$.

Applying $(\#)$ to $a_j = |t_j|^p / \sum_{j=1}^{n} |t_j|^p, b_j = |s_j|^q / \sum_{j=1}^{n} |s_j|^q$ gives

$$|t_j s_j| / (\sum_{j=1}^{n} |t_j|^p)^{1/p}(\sum_{j=1}^{n} |s_j|^q)^{1/q} \le \frac{a_j}{p} + \frac{b_j}{q}.$$

Adding these inequalities gives

$$\sum_{j=1}^{n} |t_j s_j| \le \sum_{j=1}^{n} (\frac{a_j}{p} + \frac{b_j}{q})[(\sum_{j=1}^{n} |t_j|^p)^{1/p}(\sum_{j=1}^{n} |s_j|^q)^{1/q}]$$
$$= (\frac{1}{p} + \frac{1}{q})(\sum_{j=1}^{n} |t_j|^p)^{1/p}(\sum_{j=1}^{n} |s_j|^q)^{1/q} = (\sum_{j=1}^{n} |t_j|^p)^{1/p}(\sum_{j=1}^{n} |s_j|^q)^{1/q}.$$

From Hölder's Inequality, we have Minkowski's inequality.

Theorem 1.13. *(Minkowski's Inequality)* If $p \ge 1$, then

$$(\sum_{j=1}^{n} |s_j + t_j|^p)^{1/p} \le (\sum_{j=1}^{n} |s_j|^p)^{1/p} + (\sum_{j=1}^{n} |t_j|^p)^{1/p} \quad for \quad s_j, t_j \in \mathbb{C}.$$

Proof: For $p = 1$, the inequality is trivial so assume $p > 1$. Clearly

$$(*) \quad (\sum_{j=1}^{n} |s_j + t_j|^p)^{1/p} \leq (\sum_{j=1}^{n} (|s_j| + |t_j|)^p)^{1/p}.$$

Now $(|s_j| + |t_j|)^p = (|s_j| + |t_j|)^{p-1} |s_j| + (|s_j| + |t_j|)^{p-1} |t_j|$, so by Hölder's inequality

$$\sum_{j=1}^{n} (|s_j| + |t_j|)^p = \sum_{j=1}^{n} (|s_j| + |t_j|)^{p-1} |s_j| + \sum_{j=1}^{n} (|s_j| + |t_j|)^{p-1} |t_j|$$

$$\leq (\sum_{j=1}^{n} |s_j|^p)^{1/p} (\sum_{j=1}^{n} (|s_j| + |t_j|)^{(p-1)q})^{1/q}$$

$$+ (\sum_{j=1}^{n} |t_j|^p)^{1/p} (\sum_{j=1}^{n} (|s_j| + |t_j|)^{(p-1)q})^{1/q}$$

$$= ((\sum_{j=1}^{n} |s_j|^p)^{1/p} + (\sum_{j=1}^{n} |t_j|^p)^{1/p})(\sum_{j=1}^{n} (|s_j| + |t_j|)^p)^{1/q}.$$

Dividing gives

$$(\sum_{j=1}^{n} (|s_j| + |t_j|)^p)^{1/p} \leq (\sum_{j=1}^{n} |s_j|^p)^{1/p} + (\sum_{j=1}^{n} |t_j|^p)^{1/p}$$

which by (*) gives the Minkowski inequality.

For $p = 2$, Minkowski's inequality is called the Cauchy-Schwarz Inequality.

From Minkowski's inequality, it follows that $\|\cdot\|_p$ defines a norm on l^p for $1 \leq p < \infty$.

Definition 1.14. A normed linear space X is called a Banach space if it is complete with respect to the metric induced by the norm.

We will show that some of the spaces defined above are Banach spaces and leave the verification of other spaces to the exercises.

Proposition 1.15. *The space $B(S)$ is a Banach space under the sup-norm.*

Proof: Let $\{f_j\}$ be a Cauchy sequence in $B(S)$. As is the case in most completeness proofs, we must first find a candidate for the limit of the sequence $\{f_j\}$, then verify that the candidate belongs to the space being considered and that the sequence converges to the candidate. For each $t \in S$, we have $|f_j(t) - f_k(t)| \leq \|f_j - f_k\|_\infty$ so $\{f_j(t)\}$ is a Cauchy sequence in the scalar field which is complete. Therefore, there exists $f(t)$ such that $\lim_j f_j(t) = f(t)$ so we have a function $f : S \to \Bbbk$. We need to show that $f \in B(S)$ and $\|f_j - f\|_\infty \to 0$. Let $\epsilon > 0$. There exists N such that $j, k \geq N$ implies that $\|f_j - f_k\|_\infty < \epsilon$. If $t \in S$ and $j, k \geq N$, $|f_j(t) - f_k(t)| \leq \|f_j - f_k\|_\infty < \epsilon$ so $|f_j(t) - f(t)| \leq \epsilon$ for $j \geq N$. Hence, $|f(t)| \leq |f_N(t)| + \epsilon$ so $f \in B(S)$ and we also have $\|f_j - f\|_\infty \leq \epsilon$ for $j \geq N$. Thus, $\{f_j\}$ converges to f.

In particular, the space l^∞ is a Banach space.

Corollary 1.16. *If S is a compact metric space, then $C(S)$ is a Banach space.*

Proof: $C(S)$ is a closed subspace of $B(S)$ (the uniform limit of continuous functions is continuous).

Example 1.17. The subspace c_{00} of l^∞ is not complete. For example, the sequence $\{\sum_{j=1}^n e^j/n\}$, where e^j is the sequence with a 1 in the jth coordinate and 0 in the other coordinates, is a Cauchy sequence in c_{00} which does not converge to an element of c_{00}. Similarly, the subspace m_0 of l^∞ is not complete (consider the same sequence).

In Exercise 3 the reader is asked to show that the subspaces c_0 and c are Banach spaces. We next show that the spaces l^p are Banach spaces for $1 \leq p < \infty$.

Example 1.18. Let $1 \leq p < \infty$ and let $x_k = \{t_{kj}\}_{j=1}^\infty$ be a Cauchy sequence in l^p. For each j, $|t_{mj} - t_{nj}| \leq \|x_m - x_n\|_p$ so $\{t_{nj}\}_{n=1}^\infty$ is a scalar Cauchy sequence and converges to some t_j. Set $x = \{t_j\}$. We show that $x \in l^p$ and $\|x_n - x\|_p \to 0$. Let $\epsilon > 0$. There exists N such that $n, m \geq N$ implies $\|x_n - x_m\|_p < \epsilon$. Then for any M, $\sum_{j=1}^M |t_{mj} - t_{nj}|^p \leq \|x_n - x_m\|_p^p < \epsilon^p$. Letting $n \to \infty$ with M and m fixed gives $\sum_{j=1}^M |t_{mj} - t_j|^p \leq \epsilon^p$ for $m \geq N$. Thus, $\sum_{j=1}^\infty |t_{mj} - t_j|^p \leq \epsilon^p$ for $m \geq N$. It follows that $\{t_{mj} - t_j\}_{j=1}^\infty \in l^p$ so $x \in l^p$ and $\|x_n - x\|_p \to 0$.

Example 1.19. Consider the space $C^1[a, b]$ of functions with one continuous derivative. The space $C^1[a, b]$ is subspace of $C[a, b]$, but is not complete with respect to the sup-norm (any continuous function is the uniform limit of polynomials by the Weierstrass Approximation Theorem ([DeS])18.8).

However, there is another natural norm under which $C^1[a,b]$ is complete, namely, the norm $\|f\|_1 = \|f\|_\infty + \|f'\|_\infty$ (this follows, for example, from Theorem 11.7 of [DeS]).

We now give a series criterion for checking completeness which is often useful.

Definition 1.20. A series $\sum_j x_j$ in a semi-normed space is absolutely convergent if $\sum_{j=1}^\infty \|x_j\| < \infty$.

Theorem 1.21. *Let X be a semi-normed space. Then X is complete iff every absolutely convergent series $\sum_j x_j$ in X is convergent.*

Proof: Suppose that X is complete and the series $\sum_j x_j$ is absolutely convergent. Let $s_n = \sum_{j=1}^n x_j$ be the nth partial sum of the series $\sum_j x_j$. Then $\{s_n\}$ is a Cauchy sequence since $\|s_{n+p} - s_n\| \le \sum_{j=n+1}^{n+p} \|x_j\|$ and, therefore, converges.

Conversely, assume that every absolutely convergent series converges. Let $\{x_j\}$ be a Cauchy sequence. It suffices to show that $\{x_j\}$ has a convergent subsequence. For each k pick an n_k such that $n \ge n_k$ implies $\|x_n - x_{n_k}\| < 1/2^k$, where we may assume that the $\{n_k\}$ are increasing. Then

$$\sum_{k=1}^\infty \|x_{n_{k+1}} - x_{n_k}\| \le \sum_{k=1}^\infty 1/2^k$$

so the series $\sum_k (x_{n_{k+1}} - x_{n_k})$ is absolutely convergent. Therefore, there exists $x \in X$ such that $\lim_N \sum_{k=1}^N (x_{n_{k+1}} - x_{n_k}) = \lim_N (x_{n_{N+1}} - x_{n_1}) = x$. Hence, $x_{n_k} \to x + x_{n_1}$ as desired.

Remark 1.22. For readers familar with the Lebesgue integral we mention the corresponding Lebesgue spaces. Let $E \subset \mathbb{R}^n$ be a Lebesgue measurable subset and $1 \le p < \infty$. The space $L^p(E)$ consists of all Lebesgue measurable functions $f : E \to \Bbbk$ such that $|f|^p$ is Lebesgue integrable over E. The Hölder and Minkowski inequalities hold so $f \to (\int_E |f|^p)^{1/p} = \|f\|_p$ defines a semi-norm on $L^p(E)$ such that $\|f\|_p = 0$ iff $f = 0$ almost everywhere in E. It is customary to identify functions which are equal almost everywhere so $L^p(E)$ then becomes a normed space which is actually a Banach space. This completeness is one of the principal reasons for employing the Lebesgue integral (see [HS],[Sw1]). See also Exercise 5.

For descriptions of the history of the evolution of functional analysis, see [Di],[Mon],[Sm] and [SS].

Exercises.

1. Let $(X, \|\cdot\|_X), (Y, \|\cdot\|_Y)$ be semi-normed spaces. Show that $X \times Y$ is a semi-normed space under the semi-norm $\|(x, y)\| = \|x\|_X + \|y\|_Y$. Give necessary and sufficient conditions for this semi-norm to be a norm.

2. Let X be a semi-normed space and fix $x_0 \in X$ and $t_0 \in \Bbbk, t_0 \neq 0$. Show that the map $x \rightarrow t_0 x + x_0$ is a continuous map from X onto X with a continuous inverse.

3. Show that the subspaces c_0 and c of l^∞ are complete. What about the subspace c_c?

4. Show that bs is a Banach space and that cs is a closed subspace.

5. Let $\mathcal{R}[a, b]$ be the space of all functions $f : [a, b] \rightarrow \mathbb{R}$ which are Riemann integrable over $[a, b]$. Show $\|f\| = \int_a^b |f|$ defines a semi-norm on $\mathcal{R}[a, b]$ which is not a norm. Does this define a norm on the subspace $C[a, b]$? (This semi-norm is not complete, but examples are not easy; see [Sw1]3.3. This lack of completeness is one of the reasons that the Lebesgue integral is considered superior to the Riemann integral).

6. Let $C^n[a, b]$ be the space of all functions $f : [a, b] \rightarrow \mathbb{R}$ with n continuous derivatives. Show that $C^n[a, b]$ is a Banach space under the norm $\|f\|_n = \sum_{j=0}^n \|f^j\|_\infty$, where $f^0 = f$.

7. Show that $BV[a, b]$ is a Banach space. (See Appendix A.)

8. Give an example of an absolutely convergent series in c_{00} which does not converge.

9. A subset B of a semi-normed space X is *bounded* if there exists $M > 0$ such that $\|x\| \leq M$ for all $x \in B$. Show that a Cauchy sequence in a semi-normed space is bounded.

10. Show that a subset B of a semi-normed space is bounded iff for every $\{x_j\} \subset B$ and scalar sequence $t_j \rightarrow 0$, $t_j x_j \rightarrow 0$.

11. If L is a linear subspace of a semi-normed space, show L^-, the closure of L, is a subspace.

12. If L is a linear subspace of a semi-normed space X such that L contains an open set, show $L = X$.

13. If a normed space X contains a sequence $\{x_j\}$ such that the subspace spanned by $\{x_j\}$ is dense in X, show X is separable. (A metric space is *separable* if it contains a countable dense subset.)

14. If $1 \leq p < \infty$, show l^p is separable. Show l^∞ is not separable (Hint: consider the subset of sequences which are either 0 or 1).

15. Are c, c_0 separable?

Chapter 2

Linear Operators

In this chapter we will develop the basic properties of continuous linear operators between semi-normed spaces. Let X, Y be semi-normed spaces over $\Bbbk$. Let $L(X, Y)$ be the space of all continuous linear operators from X into Y. Then $L(X, Y)$ is a vector space under the usual operations of pointwise addition and scalar multiplication. If $X = Y$, we write $L(X, X) = L(X)$, and if $Y = \Bbbk$, we write $L(X, \Bbbk) = X'$ and X' is called the *dual space* of X.

Theorem 2.1. *Let $T : X \to Y$ be linear. The following are equivalent: (i) T is continuous on X, (ii) T is continuous at 0, (iii) $\{\|Tx\| : \|x\| \leq 1\}$ is bounded in $\mathbb{R}$, (iv) there exists $M > 0$ such that $\|Tx\| \leq M \|x\|$ for all $x \in X$. [Note that we do not distinguish between the semi-norms on X and Y; it will be clear from the context on which spaces the semi-norms are operating.]*

Proof: Clearly (i) implies (ii). Assume (ii) and that (iii) does not hold. Then there exists a sequence $\{x_j\} \subset X$ with $\|x_j\| \leq 1$ and $\|Tx_j\| > j$. Set $y_j = x_j/j$ so $y_j \to 0$ but $\|Ty_j\| = \frac{1}{j} \|Tx_j\| > 1$. This implies that T is not continuous at 0. Hence, (ii) implies (iii).

Assume (iii) holds. By (iii) there exists $M > 0$ such that $\|Tx\| \leq M$ for $\|x\| \leq 1$. Let $x \in X$. If $x = 0$, (iv) holds trivially so assume $x \neq 0$. Then $\|T(x/\|x\|)\| \leq M$ or $\|Tx\| \leq M \|x\|$ so (iv) holds.

Assume (iv) holds. If $x, y \in X$, then $\|Tx - Ty\| = \|T(x - y)\| \leq M \|x - y\|$ which implies that T is (uniformly) continuous.

From (iii) it follows that a linear operator is continuous iff it carries bounded sets into bounded sets (see Exercise 1.9); for this reason continuous linear operators are often referred to as *bounded operators*.

9

Definition 2.2. Let $T \in L(X, Y)$. The operator semi-norm of T is defined to be $\|T\| = \sup\{\|Tx\| : \|x\| \leq 1\}$. [Note again we do not distinguish between the semi-norms on the various spaces.]

See Exercise 1 for other expressions for the operator semi-norm.

Note that if $T \in L(X, Y)$, then $\|Tx\| \leq \|T\| \|x\|$ for all $x \in X$.

Theorem 2.3. *$L(X, Y)$ is a semi-normed space under the operator semi-norm and is a normed space if Y is a normed space. If Y is a Banach space, then $L(X, Y)$ is a Banach space.*

Proof: Let $T \in L(X, Y)$. Clearly $\|T\| \geq 0$ and $\|tT\| = |t| \|T\|$. If Y is a normed space, then $\|T\| = 0$ iff $Tx = 0$ for all x or iff $T = 0$.

For the triangle inequality, let $S \in L(X, Y)$. Then

$$\|(T + S)x\| \leq \|Tx\| + \|Sx\| \leq \|x\| (\|T\| + \|S\|)$$

which implies that $\|T + S\| \leq \|T\| + \|S\|$.

Now suppose that Y is a Banach space and let $\{T_j\}$ be a Cauchy sequence with respect to the operator norm. Then for every $x \in X$, $\{T_j x\}$ is a Cauchy sequence in Y since $\|T_j x - T_k x\| \leq \|T_j - T_k\| \|x\|$. Let $Tx = \lim_j T_j x$. Then T defines a linear map from X into Y (Proposition 1.2). Since $\{T_j\}$ is Cauchy, there exists $M > 0$ such that $\|T_j\| \leq M$ for all j (Exercise 1.9). Therefore, $\|T_j x\| \leq M \|x\|$ for all x and $\|Tx\| \leq M \|x\|$ by Proposition 1.3. Hence, $T \in L(X, Y)$ by Theorem 1.

We show $\|T_j - T\| \to 0$. Let $\epsilon > 0$. There exists N such that $j, k \geq N$ implies $\|T_j - T_k\| < \epsilon$. If $\|x\| \leq 1$ and $j, k \geq N$, then $\|T_j x - T_k x\| \leq \|T_j - T_k\| < \epsilon$. Letting $k \to \infty$ gives $\|T_j x - Tx\| \leq \epsilon$ (Proposition 1.3) or $\|T_j - T\| \leq \epsilon$ for $j \geq N$.

Corollary 2.4. *The dual space X' is always complete under the norm, $\|x'\| = \sup\{|x'(x)| : \|x\| \leq 1\}$, $x' \in X'$, this norm is called the dual norm on X'.*

Later, we will use the Hahn-Banach Theorem to show that the converse of the last statement in Theorem 3 holds. We will give examples of the dual spaces of some of the normed spaces given in the examples of Chapter 1 at the end of this chapter.

Proposition 2.5. *Let Z be a semi-normed space, $T \in L(X, Y), S \in L(Y, Z)$, Then $ST \in L(X, Z)$ and $\|ST\| \leq \|S\| \|T\|$. [Here ST denotes the composition of S and T.]*

Proof: Let $\|x\| \leq 1$. Then $\|STx\| \leq \|S\| \|Tx\| \leq \|S\| \|T\| \|x\|$ so the result follows.

We now give several examples of continuous linear operators.

Example 2.6. (Volterra operator) Define $V : C[0,1] \to C[0,1]$ by $Vf(t) = \int_0^t f$. Then V is continuous and clearly $\|V\| \leq 1$. Since $\|V(1)\| = \|t\| = 1$, it follows that $\|V\| = 1$.

Example 2.7. Let $k \in C([a,b] \times [a,b])$ and define $K : C[a,b] \to C[a,b]$ by $Kf(t) = \int_a^b k(t,s)f(s)ds$. Then $\|Kf\|_\infty \leq \|k\|_\infty \|f\|_\infty (b-a)$ so K is continuous with $\|K\| \leq \|k\|_\infty (b-a)$. K is called an integral operator induced by the kernel k.

Example 2.8. Let $X = C^1[0,1]$ and let D be the differential operator $D : X \to C[0,1]$ defined by $Df = f'$. Then D is not continuous with respect to the sup-norm since if $f_k(t) = t^k$, then $\|Df_k\|_\infty = k$ while $\|f_k\|_\infty = 1$. However, if X is supplied with the norm $\|\cdot\|_1$ of Example 1.7, then D is continuous since $\|Df\|_\infty \leq \|f\|_1$.

If $A = [a_{ij}]$ is an infinite matrix and E and F are vector spaces of sequences, we say that A maps E into F if for each $t = \{t_j\} \in E$ the series $\sum_{j=1}^\infty a_{ij}t_j$ converges for every $i \in \mathbb{N}$ and the sequence $At = \{\sum_{j=1}^\infty a_{ij}t_j\} \in F$. One of the principal problems in this area is to give characterizations of matrices which map one sequence space into another; characterizations of matrices mapping c_0 into itself are given in Chapter 12 and those mapping l^1 into l^p ($1 \leq p < \infty$) in Chapter 13. Another interesting problem in this area is what is referred to as automatic continuity; if a matrix A maps a normed sequence space E into another normed sequence space F and the spaces satisfy appropriate conditions, the matrix map is then (automatically) continuous. We give examples of such automatic continuity results in 8.8, 12.12 and 19.3. We now give an example of a continuous matrix operator.

Example 2.9. (Summation operator) Let S be the matrix $S = [s_{ij}]$ defined by $s_{ij} = 1$ if $i \geq j$ and $s_{ij} = 0$ otherwise. Then $S : l^1 \to l^\infty$ is defined by $S(\{t_j\}) = \{\sum_{i=1}^j t_i\}$ and S is continuous with $\|S\| \leq 1$.

Other examples are given in the exercises.

Theorem 2.10. *Let Z be a Banach space and X a dense linear subspace of Y. If $T \in L(X,Z)$, then T has a unique continuous linear extension $T^\# : Y \to Z$ with $\|T\| = \|T^\#\|$.*

Proof: Let $y \in X$ and $x_j \in X$ with $x_j \to y$. Then $\{Tx_j\}$ is Cauchy in Z since $\|Tx_j - Tx_k\| \le \|T\| \, \|x_j - x_k\|$. Therefore, $\{Tx_j\}$ converges to some $T_1 y \in Z$. If, also, $u_j \in X$ converges to y, then $\{Tu_j\}$ converges to some $T_2 y \in Z$. Since $\|T_1 y - T_2 y\| = \lim \|Tx_j - Tu_j\| \le \|T\| \lim \|x_j - u_j\| = 0$, $T_1 y = T_2 y$. That is, the limit, $\lim Tx_j$, is independent of the sequence converging to y. Therefore, we may define a linear map $T^{\#} : Y \to Z$ by $T^{\#} y = \lim Tx_j$ which extends T. Since $\|T^{\#} y\| = \lim \|Tx_j\| \le \|T\| \lim \|x_j\| = \|T\| \, \|y\|$, $T^{\#}$ is continuous with $\|T^{\#}\| \le \|T\|$. Clearly $\|T\| \le \|T^{\#}\|$ so $\|T\| = \|T^{\#}\|$.

Completeness in Theorem 10 cannot be dropped; see Exercise 15. We now establish several results concerning the inverse of continuous linear operators. In what follows $\mathcal{R}A$ will denote the range of the operator A, I will denote the identity operator and $A^0 = I$.

Theorem 2.11. *Let $T \in L(X)$. Then T^{-1} exists as a continuous linear operator on $\mathcal{R}T$ iff there exists $m > 0$ such that $\|Tx\| \ge m \, \|x\|$ for all $x \in X$.*

Proof: Suppose T^{-1} exists and is continuous (on $\mathcal{R}T$) and set $m = 1/\|T^{-1}\|$. Then $\|T^{-1}(Tx)\| = \|x\| \le \|T^{-1}\| \, \|Tx\|$.

Conversely, assume the inequality holds. Then T is one-one so T^{-1} exists and $\|T^{-1}(Tx)\| = \|x\| \le (1/m) \|Tx\|$ implies that T^{-1} is continuous on $\mathcal{R}T$.

Theorem 2.12. *Let X be a normed space and suppose $A \in L(X)$. If the series $\sum_{j=0}^{\infty} A^j$ converges in norm in $L(X)$, then $I - A$ is invertible with $(I - A)^{-1} = \sum_{j=0}^{\infty} A^j$.*

Proof: If $S = \sum_{j=0}^{\infty} A^j$, then $SA = AS = \sum_{j=0}^{\infty} A^{j+1}$ so $(I - A)S = S(I - A) = I$.

The series $\sum_{j=0}^{\infty} A^j$ is called the *Neumann series* for $(I - A)^{-1}$. If X is a Banach space, a sufficient condition for the convergence of the series is $\|A\| < 1$ since in this case we have $\|A^j\| \le \|A\|^j$ and the series $\sum_{j=0}^{\infty} A^j$ converges absolutely and , therefore, converges in $L(X)$ (Theorems 3 and 1.21). In this case, we also have $\|(I - A)^{-1}\| \le 1/(1 - \|A\|)$.

We can improve the sufficient condition that $\|A\| < 1$ for the convergence of the Neumann series. We show this condition can be replaced by the sufficient condition $\lim \sqrt[j]{\|A^j\|} < 1$.

Lemma 2.13. *Let $a_j \in \mathbb{R}$ satisfy $0 \le a_{j+k} \le a_j a_k$ for all $j, k \in \mathbb{N}$. Then $\{\sqrt[j]{a_j}\}$ converges to $a = \inf\{\sqrt[j]{a_j} : j \in \mathbb{N}\}$.*

Proof: Let $\epsilon > 0$. Choose j such that $\sqrt[j]{a_j} < a + \epsilon$. For $k > j$, write $k = qj + r$, where $0 \le r \le j - 1$. Set $M = \max\{a_r : 0 \le r \le j - 1\}$. Then $a_k \le a_j...a_j a_r = (a_j)^q a_r$ so $a \le \sqrt[k]{a_k} \le \sqrt[k]{a_r}(a_j^{1/j})^{\frac{qj}{qj+r}} \le \sqrt[k]{M}(a+\epsilon)^{\frac{qj}{qj+r}} \to a + \epsilon$ as $k \to \infty$ and the result follows.

Corollary 2.14. *Let $A \in L(X)$. Then $\lim \sqrt[j]{\|A^j\|}$ exists and $\lim \sqrt[j]{\|A^j\|} \le \|A\|$.*

Proof: Set $a_j = \|A^j\|$ in Lemma 13.

The operator in Exercise 17 show that strict inequality can occur in Corollary 14.

The root test from elementary calculus ([Ap]8.26,[DeS]3.14) carries forward to determine absolute convergence for series of operators.

Theorem 2.15. *(Root Test). Let $T_j \in L(X)$ and set $r = \overline{\lim} \sqrt[j]{\|T_j\|}$. If $r < 1$, then the series $\sum_{j=1}^{\infty} T_j$ converges absolutely and if $r > 1$, then the series $\sum_{j=1}^{\infty} T_j$ does not converge (in fact, $\|T_j\| \nrightarrow 0$).*

From Theorems 12 and 15 we have a sufficient condition for the convergence of the Neumann series.

Corollary 2.16. *Let X be a Banach space and $A \in L(X)$. If $\lim \sqrt[j]{\|A^j\|} < 1$, the Neumann series $\sum_{j=0}^{\infty} A^j$ converges (absolutely) in norm to $(I - A)^{-1}$.*

The operator in Exercise 17 shows that the condition in Corollary 16 improves the sufficient condition $\|A\| < 1$.

Theorem 2.17. *Let X be a Banach space and $A \in L(X)$ be such that $\mathcal{R}A = X$, A^{-1} exists and belongs to $L(X)$. If $B \in L(X)$ is such that $\|A - B\| < 1/\|A^{-1}\|$, then $\mathcal{R}B = X$, B^{-1} exists and is in $L(X)$ with*

$$\|B^{-1}\| \le \|A^{-1}\|/(1 - \|A^{-1}\|\|A - B\|)$$

and

$$\|B^{-1} - A^{-1}\| \le \|A^{-1}\|^2 \|A - B\|/(1 - \|A^{-1}\|\|A - B\|).$$

Proof: Now $B = A - (A - B) = A(I - A^{-1}(A - B))$. Since $\|A^{-1}(A - B)\| \le \|A^{-1}\|\|A - B\| < 1$, by the remarks about Neumann series following Theorem 12, $(I - A^{-1}(A - B))^{-1} = \sum_{j=0}^{\infty}[A^{-1}(A - B)]^j$. Thus, $B^{-1} \in L(X)$ and $B^{-1} = (I - A^{-1}(A - B))^{-1}A^{-1}$. Also, $B^{-1} = \sum_{j=0}^{\infty}[A^{-1}(A - B)]^j A^{-1}$ so the first inequality in the statement of the theorem is clear. For the second inequality in the statement, we have

$$\|B^{-1} - A^{-1}\| = \|(B^{-1}A - I)A^{-1}\| \leq \|A^{-1}\| \|B^{-1}A - I\|$$
$$= \|A^{-1}\| \left\|I - \sum_{j=0}^{\infty}[A^{-1}(A - B)]^j\right\| = \|A^{-1}\| \left\|\sum_{j=1}^{\infty}[A^{-1}(A - B)]^j\right\|$$
$$\leq \|A^{-1}\|^2 \|A - B\| / (1 - \|A^{-1}\| \|A - B\|).$$

From the statements and inequalities in Theorem 17, we have

Corollary 2.18. *Let X be a Banach space and $\mathcal{G} = \{A \in L(X) : A^{-1} \in L(X)\}$. Then $\mathcal{G}$ is open in $L(X)$ and the map $A \to A^{-1}$ is continuous from $\mathcal{G}$ onto $\mathcal{G}$.*

We now give an application of the results above to linear integral equations. Let $X = C[a, b]$ and $k \in C([a, b] \times [a, b])$ and let K be the integral operator on X induced by k as in Example 7. Suppose that $\|k\|_{\infty}(b-a) < 1$ so $\|K\| < 1$.

Consider the Fredholm integral equation

$$(*) \quad y(t) = x(t) - \int_a^b k(t, s)x(s)ds$$

or as operators on X, $y = (I - K)x$. Since $\|K\| < 1$, $I - K$ has an inverse so $x = (I - K)^{-1}y$ is the solution to $(*)$ and $x = \sum_{j=0}^{\infty} K^j y$.

We now give a description of the K^j as integral operators, i.e., we compute the kernels of the K^j. Define kernels k_j by $k_1(s, t) = k(s, t)$ and $k_{j+1}(s, t) = \int_a^b k_1(s, u)k_j(u, t)du$ for $j > 1$. Now

$$K^2 y(s) = \int_a^b k(s, t) \int_a^b y(u)k(t, u)du\,dt$$
$$= \int_a^b y(u) \int_a^b k(s, t)k(t, u)dt\,du = \int_a^b y(u)k_2(s, u)du.$$

Inductively, $K^j y(s) = \int_a^b k_j(s, t)y(t)dt$ for $j \geq 1$. That is, the operator K^j is the integral operator induced by the kernel k_j. Therefore,

$$(\&) \quad x(s) = \sum_{j=0}^{\infty} K^j y(s) = y(s) + \sum_{j=1}^{\infty} \int_a^b k_j(s, t)y(t)dt,$$

where the series converges uniformly for $s \in [a, b]$.

Consider the series

$$(\#) \quad h(s,t) = \sum_{j=1}^{\infty} k_j(s,t).$$

Inductively, $|k_j(s,t)| \leq M^j(b-a)^{j-1}$ $(j > 1)$, where $M = \|k\|_\infty$. Since $M(b-a) < 1$, $(\#)$ converges uniformly so $(\&)$ becomes

$$x(s) = y(s) + \int_a^b (\sum_{j=1}^{\infty} k_j(s,t))y(t)dt = y(s) + \int_a^b h(s,t)y(t)dt.$$

Or, in operator notation $x = (I+H)y$, where H is the integral operator induced by the kernel h, so that $(I+K)^{-1} = I+H$. The kernel h is called the *reciprocal kernel*.

We now give several examples of dual spaces for the spaces described in the examples of Chapter 1. If X and Y are normed spaces, a linear operator $U : X \rightarrow Y$ is called an *isometry* if $\|Ux\| = \|x\|$ for all $x \in X$. If an isometry U is onto Y, then the spaces are isometrically isomorphic and we identify X and Y and write $X = Y$.

Theorem 2.19. *Let* $t = \{t_j\} \in l^1$. *Then* t *induces a continuous linear functional* $F_t : c_0 \rightarrow \mathbb{R}$ *defined by* $F_t(s) = F_t(\{s_j\}) = \sum_{j=1}^{\infty} t_j s_j$ *with* $\|F_t\| = \|t\|_1$. *The linear map* $U : l^1 \rightarrow (c_0)'$, $Ut = F_t$, *is an isometry onto* $(c_0)'$ *so* $l^1 = (c_0)'$.

Proof: If $s \in c_0$, then $|F_t(s)| \leq \|t\|_1 \|s\|_\infty$ so F_t is linear and continuous with $\|F_t\| \leq \|t\|_1$. Let e^j be the sequence with 1 in the jth coordinate and 0 in the other coordinates. Then

$$\left| F_t(\sum_{j=1}^{n}(sign\, t_j)e^j) \right| = \sum_{j=1}^{n} |t_j| \leq \|F_t\| \left\| \sum_{j=1}^{n}(sign\, t_j)e^j \right\| = \|F_t\|$$

so $\|t\|_1 \leq \|F_t\|$ and $\|t\|_1 = \|F_t\|$. Thus, the linear map $U : l^1 \rightarrow (c_0)'$, $Ut = F_t$, is an isometry. We show that U is onto. Let $F \in (c_0)'$ and set $t_j = F(e^j)$, $t = \{t_j\}$. If $s = \{s_j\} \in c_0$, then $s = \lim_n \sum_{j=1}^{n} s_j e^j$ so $F(s) = \sum_{j=1}^{\infty} t_j s_j$. Also,

$$\left| F(\sum_{j=1}^{n}(sign\, t_j)e^j) \right| = \sum_{j=1}^{n} |t_j| \leq \|F\| \left\| \sum_{j=1}^{n}(sign\, t_j)e^j \right\| = \|F\|$$

implies $t \in l^1$ and we have $Ut = F$.

Theorem 2.20. *Let* $t = \{t_j\} \in l^\infty$. *Then* t *induces a continuous linear functional* $F_t : l^1 \rightarrow \mathbb{R}$ *defined by* $F_t(s) = F_t(\{s_j\}) = \sum_{j=1}^{\infty} t_j s_j$ *with*

$\|F_t\| = \|t\|_\infty$. *The linear map* $U : l^\infty \to (l^1)', Ut = F_t$, *is an isometry onto* $(l^1)'$ *so* $l^\infty = (l^1)'$.

Proof: If $s \in l^1$, then $|F_t(s)| \le \|s\|_1 \|t\|_\infty$ so F_t is linear and continuous with $\|F_t\| \le \|t\|_\infty$. Now $|F_t(e^j)| = |t_j| \le \|F_t\|$ so $\|t\|_\infty \le \|F_t\|$ and $\|t\|_\infty = \|F_t\|$: Thus, the linear map $U : l^\infty \to (l^1)', Ut = .F_t$, is an isometry. We show that U is onto. Let $F \in (l^1)'$ and set $t_j = F(e^j), t = \{t_j\}$. Since $|F(e^j)| \le \|F\|$, $t \in l^\infty$ and we have $Ut = F$.

We have shown that the dual of c_0 is l^1 and the dual of l^1 is l^∞ but the dual of l^∞ is much more complicated and difficult to describe (see [Sw1]6.3,[DuS]).

Theorem 2.21. *Let* $1 < p < \infty$ *and* $\frac{1}{p} + \frac{1}{q} = 1$. *Let* $t = \{t_j\} \in l^q$. *Then* t *induces a continuous linear functional* $F_t : l^p \to \mathbb{R}$ *defined by* $F_t(s) = F_t(\{s_j\}) = \sum_{j=1}^\infty t_j s_j$ *with* $\|F_t\| = \|t\|_q$. *The linear map* $U : l^q \to (l^p)', Ut = F_t$, *is an isometry onto* $(l^p)'$ *so* $l^q = (l^p)'$.

Proof: By the Hölder inequality, $|F_t(s)| \le \|t\|_q \|s\|_p$ so F_t is linear and continuous with $\|F_t\| \le \|t\|_q$. Set $x^n = \sum_{j=1}^n |t_j|^{q-1} (sign t_j) e^j = \{x_j^n\}_j$. Then $F_t(x^n) = \sum_{j=1}^n |t_j|^q \le \|F_t\| \|x^n\|_p = \|F_t\| (\sum_{j=1}^n |t_j|^q)^{1/p}$ so $(\sum_{j=1}^n |t_j|^q)^{1-1/p} = (\sum_{j=1}^n |t_j|^q)^{1/q} \le \|F_t\|$ and $(\sum_{j=1}^\infty |t_j|^q)^{1/q} = \|t\|_q \le \|F_t\|$. Hence, $\|F_t\| = \|t\|_q$ so the linear map $U : l^q \to (l^p)', Ut = F_t$, is an isometry. We show that U is onto. Let $F \in (l^p)'$ and set $t_j = F(e^j), t = \{t_j\}$. Now as above $F(x^n) = \sum_{j=1}^n |t_j|^q \le \|F\| \|x^n\|_p = \|F\| (\sum_{j=1}^n |t_j|^q)^{1/p}$ and $(\sum_{j=1}^n |t_j|^q)^{1/q} \le \|F\|$ so $t \in l^q$ and we have $Ut = F$.

We give examples of the duals of other spaces in the exercises. The dual of $C[a, b]$ which involves functions of bounded variation and the Riemann-Stieltjes integral is described in Appendix C.

Remark 2.22. As in Chapter 1 we offer some remarks for the reader familiar with the Lebesgue integral. As in Theorems 20 and 21 the dual of $(L^p(E), \|\cdot\|_p)$ is $(L^q(E), \|\cdot\|_q)$ with the correspondence between continuous linear functions F on $(L^p(E), \|\cdot\|_p)$ being given by the existence of a function f belonging to $(L^q(E), \|\cdot\|_q)$ with $F(g) = \int_E fg$, for all $g \in L^p(E), \|F\| = \|f\|_p$. The case $p = 1$ is more complicated. The space $L^\infty(E)$ consists of all measurable functions $f : E \to \Bbbk$ such that there exists $M \ge 0$ such that $|f(t)| \le M$ for all t except those belonging to a subset of measure 0. A semi-norm $\|\cdot\|_\infty$ is defined on $L^\infty(E)$ by

$\|f\|_\infty = \inf\{\sup\{|f(t)| : t \in E \backslash N\} : N$ with measure $0\}$. Again $\|f\|_\infty = 0$ iff $f = 0$ almost everywhere so it is customary to identify functions which are equal almost everywhere so $(L^\infty(E), \|\cdot\|_\infty)$ is a normed space which is actually a Banach space. The dual of $L^1(E)$ is then $(L^\infty(E), \|\cdot\|_\infty)$ with the correspondence being given by $F(g) = \int_E fg$ as above. See [Ba] or [Sw1] for details.

Exercises.

1. If $T \in L(X, Y)$, show $\|T\| = \sup\{\|Tx\| : \|x\| = 1\} = \inf\{M > 0 : \|Tx\| \leq M \|x\|, x \in X\}$.

2. Let $\|\cdot\|_2$ be the euclidean norm on $\mathbb{R}^n$. Show the dual of $\mathbb{R}^n$ with $\|\cdot\|_2$ is $(\mathbb{R}^n, \|\cdot\|_2)$.

3. If Z is a dense linear subspace of a normed space X, show that $Z' = X'$. Describe the dual of c_{00}.

4. Describe the dual of c_c.

5. Show that the dual of c is isometrically isomorphic to l^1 under the mapping $U : l^1 \to c'$, $Ut = \sum_{j=1}^\infty t_j s_{j-1}$, where $s_0 = \lim s_j, s = \{s_j\} \in c$. Note that this pairing of c' and l^1 is different than that of c_0' and l^1 given in Theorem 19.

6. Let $C = [c_{ij}]$ be the Cesaro matrix defined by $c_{ij} = 1/i$ for $1 \leq i \leq j$ and $c_{ij} = 0$ otherwise. Show that $C : l^\infty \to l^\infty$ and C is continuous with $\|C\| = 1$. Show that $Ct \in c$ and $\lim t = \lim Ct$ when $t \in c$ (Hint: $\left|\sum_{j=1}^n t_j/n - t\right| = \left|\sum_{j=1}^n (t_j - t)/n\right| \leq \left|\sum_{j=1}^N (t_j - t)/n\right| + \left|\sum_{j=N+1}^n (t_j - t)/n\right|$): Show that $C(bs) = c_0$.

7. Show that an infinite matrix mapping $A = [a_{ij}] : l^\infty \to l^\infty$ is continuous iff $\sup\{\sum_{j=1}^\infty |a_{ij}| : i \in \mathbb{N}\} = \|A\| < \infty$. Give a characterization of the matrices which map c_0 continuously into c_0.

8. If $t = \{t_j\} \in l^1$ show that $s = \{s_j\} \to \sum_{j=1}^\infty s_j t_j$ defines a continuous linear functional on l^∞ with norm $\|t\|_1$.

9. Let $1 \leq p \leq \infty$. Define the right (left) shift R (L) on l^p by $R(\{t_j\}) = (0, t_1, t_2, ...)$ $(L(\{t_j\}) = (t_2, t_3, ...))$. Show R and L are linear and continuous and compute $\|R\|, \|L\|$.

10. Show the linear functional $\lim : c \to c$ on c is continuous and compute $\|\lim\|$.

11. Let $S \neq \varnothing$ and $t \in S$. Define $\delta_t : B(S) \to \mathbb{R}$ by $\delta_t(f) = f(t)$. Show δ_t is a continuous linear functional and compute $\|\delta_t\|$.

12. Show that c and c_0 are linearly homeomorphic.

13. Let X be a normed space over $\mathbb{R}$. Show that X and $L(\mathbb{R}, X)$ are linearly isometric.

14. Let X be a Banach space and let $T \in L(X)$. Define $e^T : X \to X$ by $e^T = \sum_{j=1}^{\infty} T^j/j!$ and show $e^T \in L(X)$ with $\|e^T\| \le e^{\|T\|}$. If T and $S \in L(X)$ commute, show $e^{T+S} = e^T e^S$. Show e^T is invertible and find its inverse.

15. Show the identity operator on c_{00} cannot be continuously extended to a linear operator on c_0.

16. Let X, Y, Z be semi-normed spaces and $B : X \times Y \to Z$ be a bilinear map. Show B is continuous iff there exists $M \ge 0$ such that $\|B(x, y)\| \le M \|x\| \|y\|$ for every $x \in X, y \in Y$.

17. Define $A : C[0, 1] \to C[0, 1]$ by $Af(t) = \int_0^1 tf(s)ds$. Show $\|A\| = 1$ and $\lim \sqrt[j]{\|A^j\|} = 1/2$.

18. If V is the Volterra operator, show $V^n f(t) = (1/(n-1)!) \int_o^t (t - s)^{n-1} f(s)ds$, $\|V^n\| \le 1/n!$ and compute $\lim \sqrt[n]{\|V^n\|}$.

19. Let $\{d_j\} \subset \mathbb{C}$ and define D on l^2 by $D\{t_j\} = \{d_j t_j\}$. Show D maps l^2 into l^2 continuously iff $\{d_j\}$ is bounded. Compute $\|D\|$ in this case. Show D has a continuous inverse iff $\inf\{|d_j| : j\} > 0$ and give a formula for the inverse.

20. Show the summation operator of Example 9 is continuous from bs into l^∞.

21. Let K, L be integral operators on $C[a, b]$ induced by the kernels k, l. Show the composition operator KL is an integral operator by finding its kernel. What about the kernel for LK?

22*. (For readers with a knowledge of the Lebesque integral) Let $I = [a, b]$. If $k \in L^2(I \times I)$, show the integral operator as defined in Example 7 maps $L^2(I)$ into itself.

23. Show that if $\sum_{i=1}^{\infty} \sum_{j=1}^{\infty} |a_{ij}|^2 < \infty$, then the matrix $A = [a_{ij}]$ maps l^2 into itself continuously.

24. Let $T : X \to Y$ be linear. Show T is continuous iff T carries Cauchy sequences to Cauchy sequences.

25. Let X be a normed space and $f : X \to \Bbbk$ be linear with a closed kernel. Show f is continuous.

26. Suppose $\{T_k\}$ are invertible in $L(X)$ and $\|T_k - T\| \to 0$. Show T is invertible iff $\{\|T_k^{-1}\|\}$ is bounded.

27. Suppose $\sup\{|a_{ij}| : i, j\} < \infty$. Show $A = [a_{ij}]$ defines a continuous linear operator from l^1 into l^∞. Does A define an operator from l^1 into l^1?

28. Suppose $B \subset X'$, the dual of a normed space X. Show B is bounded iff $\sup\{|f(x_k)| : f \in B, k \in \mathbb{N}\} < \infty$ for every null sequence $\{x_k\} \subset X$.

29. Let $g \in C[a,b]$ and define $G : C[a,b] \to C[a,b]$ by $Gf = gf$. Show G is linear and continuous and find $\|G\|$.

30. Use the formulas following Corollary 18 to find the reciprocal kernel and the solution to the Fredholm integral equation $y(t) = x(t) - \lambda \int_0^1 e^{\lambda(t-s)} x(s) ds$.

Chapter 3

Quotient Spaces

Let X be a semi-normed space with M a linear subspace. As usual denote the quotient space by X/M and denote the coset $x + M = [x]$ for $x \in X$. We define a (quotient) semi-norm on X/M by $\|[x]\|' = \inf\{\|x + m\| : m \in M\}$ =distance(x, M).

Proposition 3.1. *(i)* $\|\cdot\|'$ *defines a semi-norm on* X/M.

(ii) $\|\cdot\|'$ is a norm iff M is closed.

(iii) The quotient map $x \to [x]$ is norm reducing (with respect to $\|\cdot\|'$), continuous and open.

(iv) X/M is complete iff X is complete.

(v) If X is complete and M is closed, then X/M is a Banach space.

Proof: (i): $\|[x + y]\|' = \inf\{\|x + y + m\| : m \in M\}$
$= \inf\{\|x + y + m_1 + m_2\| : m_i \in M\} \leq \inf\{\|x + m_1\| : m_1 \in M\} + \inf\{\|y + m_2\| : m_2 \in M\}$
$= \|[x]\|' + \|[y]\|'$.
$\|t[x]\|' = \inf\{\|tx + m\| : m \in M\} = |t| \inf\{\|x + m/t\| : m \in M\} = |t| \|[x]\|'$ if $t \neq 0$ and $\|t[x]\|' = 0$ if $t = 0$.

(ii): $\|[x]\|' = 0$ iff $x \in M^-$, the closure of M.

(iii): Clearly $\|x\| \geq \|[x]\|'$ so the quotient map is norm reducing and continuous. To show that the quotient map is open we show that $\{x \in X : \|x\| < 1\}$ is mapped onto $\{[x] : \|[x]\|' < 1\}$. Let $\|[x]\|' = 1 - \delta < 1$. There exists $m \in M$ such that $\|x + m\| < 1$ and $[x + m] = [x]$.

(iv): Let $\sum_j [x_j]$ be absolutely convergent in X/M with respect to $\|\cdot\|'$. For each j choose $m_j \in M$ such that $\|x_j + m_j\| \leq \|[x_j]\|' + 1/2^j$. Thus, $\sum_j (x_j + m_j)$ is absolutely convergent in X and, therefore, convergent by Theorem 1.21. Set $z = \sum_{j=1}^{\infty} (x_j + m_j)$. Then $[z] = \sum_{j=1}^{\infty} [x_j + m_j] =$

$\sum_{j=1}^{\infty}[x_j]$ by (iii).

(v): This follows from (iv) and (ii).

Let $k(X) = \{x \in X : \|x\| = 0\}$.

Proposition 3.2. *(i) $k(X)$ is a closed subspace of X.*

(ii) The quotient map $X \to X/k(X)$ is norm preserving (i.e., is an isometry).

(iii) $X/k(X)$ is a normed space and is a Banach space iff X is complete.

Proof: (i): If $x, y \in k(X)$, then $\|x + y\| \le \|x\| + \|y\| = 0$ and $\|tx\| = |t|\,\|x\| = 0$ so $k(X)$ is a subspace. $k(X)$ is closed by Proposition 1.3.

(ii): $\|x\| \ge \|[x]\|'$ by Proposition 1(iii). For $m \in k(X)$, $\|x\| = \|x\| - \|m\| \le \|x + m\| \le \|x\|$ which implies $\|x\| = \|[x]\|'$.

(iii): This follows from Proposition 1.

Definition 3.3. If $K \subset X$, the annihilator of K (in X') is defined to be $K^0 = \{x' \in X' : x'(k) = 0 \text{ for all } k \in K\}$.

Proposition 3.4. *If M is a closed subspace of X, then $(X/M)'$ and M^0 are isometrically isomorphic via the map $V : (X/M)' \to M^0$ defined by $(Vz')x = z'[x]$, $z' \in (X/M)'$.*

Proof: For $z' \in (X/M)'$, $|(Vz')x| = |z'[x]| \le \|z'\|\,\|[x]\|',x \in X$, and $(Vz')m = z'[m] = z'[0] = 0$, $m \in M$. Thus, $Vz' \in M^0$ with $\|Vz'\| \le \|z'\|$. Since $|z'[x]| = |(Vz')y| \le \|Vz'\|\,\|y\|$ for $y \in [x]$, we have $|z'[x]| \le \|Vz'\|\,\|[x]\|'$ so $\|z'\| \le \|Vz'\|$. Thus, V is an isometry.

Given $x' \in M^0$, let z' be the linear functional on X/M defined by $z'[x] = x'(x)$. Since $|z'[x]| = |x'(y)| \le \|x'\|\,\|y\|$ for $y \in [x]$, we have $|z'[x]| \le \|x'\|\,\|x\|$. Hence, $z' \in (X/M)'$ and $Vz' = x'$ so V is an isometry onto M^0.

Exercises.

1. If X, Y are semi-normed spaces and $T \in L(X, Y)$, show the induced map $\widehat{T} : X/\ker(T) \to Y$ is linear, continuous and $\|T\| = \|\widehat{T}\|$.

2. If $K \subset X$, show that K^0 is a closed subspace of X'.

3. Let M be a linear subspace of X. Show X'/M^0 and M' are isometrically isomorphic under the map $U : X'/M^0 \to M'$ defined by $U[x'] = x'_M$, where x'_M is the restriction of x' to M.

4. Show that a subset $B \subset X/M$ is bounded iff there exists a bounded subset $A \subset X$ such that $B \subset [A]$ (See Exercise 1.9).

Chapter 4

Finite Dimensional Normed Spaces

Let X, Y be normed spaces over the same field $\Bbbk$. We say that X and Y are *isomorphic* if there exists a continuous linear map from X onto Y which has a continuous inverse.

Theorem 4.1. *If X is finite dimensional with dimension n, then X is isomorphic to $\Bbbk^n$ with the euclidean norm.*

Proof: We consider the case when $\Bbbk = \mathbb{R}$; the other case is similar. Let $\{x_1, ..., x_n\}$ be a basis for X and define a linear map $T : \mathbb{R}^n \to X$ by $T(t_1, ..., t_n) = \sum_{j=1}^{n} t_j x_j$. Then T is continuous since

$$\|T(t_1, ..., t_n)\| \leq \sum_{j=1}^{n} |t_j| \, \|x_j\| \leq (\sum_{j=1}^{n} |t_j|^2)^{1/2} (\sum_{j=1}^{n} \|x_j\|^2)^{1/2}$$

by the Cauchy-Schwarz inequality. Since $\{x_1, ..., x_n\}$ is a basis, T is onto.

It remains to show that T has a continuous inverse. Let $S = \{t = (t_1, ..., t_n) \in \mathbb{R}^n : \|t\|_2 = 1\}$. Define $f : S \to \mathbb{R}$ by $f(t) = \|Tt\|$. Then f is continuous by Proposition 1.3 since T is continuous. Since S is compact, f attains its minimum at some point $s \in S$. Moreover, $f(s) > 0$ since if $f(s) = 0$, then $Ts = \sum_{j=1}^{n} s_j x_j = 0$ and $s_j = 0$ for $1 \leq j \leq n$ so $s \notin S$. Let $0 \neq t \in \mathbb{R}^n$. Then $t/\|t\|_2 \in S$ so $\|Tt\| / \|t\|_2 \geq \|Ts\| > 0$. Hence, T has a continuous inverse by Theorem 2.11.

Corollary 4.2. *Let X be a finite dimensional normed space. Then X is a Banach space and closed, bounded subsets of X are compact.*

Corollary 4.3. *Any finite dimensional subspace of a normed linear space is closed.*

Definition 4.4. Let Y be a vector space over $\Bbbk$. Two norms $\|\cdot\|_1, \|\cdot\|_2$ on Y are equivalent if there exist $a, b > 0$ such that $a \|x\|_1 \leq \|x\|_2 \leq b \|x\|_1$.

That is, $\|\cdot\|_1, \|\cdot\|_2$ on Y are equivalent iff the identity operator I : $(Y, \|\cdot\|_1) \rightarrow (Y, \|\cdot\|_2)$ is continuous and has a continuous inverse (Theorem 2.11).

Theorem 4.5. *Let X be a finite dimensional vector space with norms $\|\cdot\|_1, \|\cdot\|_2$. Then $\|\cdot\|_1$ and $\|\cdot\|_2$ are equivalent.*

Proof: Assume that $\Bbbk = \mathbb{R}$. Let T_i be the map defined in Theorem 1 from $\mathbb{R}^n \rightarrow (X, \|\cdot\|_i)$. Then each T_i is an isomorphism and since $T_1 T_2^{-1} = I$, the identity I is an isomorphism from $(X, \|\cdot\|_2)$ onto $(X, \|\cdot\|_1)$. From 2.1 and 2.11, $(1/\|I^{-1}\|) \|x\|_2 \leq \|x\|_1 \leq \|I\| \|x\|_2$.

We next consider the converse of Corollary 2. The proof utilizes a lemma of F. Riesz which is of interest in its own right and will be used later when we consider compact operators between normed spaces.

Lemma 4.6. *(F.Riesz) Let X be a normed linear space and X_0 a proper, closed subspace of X. Then for every $\theta > 0$, $0 < \theta < 1$, there exists $x_\theta \in X, \|x_\theta\| = 1$, with $\|x_\theta - x\| \geq \theta$ for $x \in X_0$.*

Proof: Let $x_1 \in X \setminus X_0$ and set $d = dist(x_1, X_0) = \inf\{\|x_1 - x\| : x \in X_0\}$. Since X_0 is closed, $d > 0$. Now there exists $x_0 \in X_0$ such that $\|x_1 - x_0\| \leq d/\theta$ since $d/\theta > d$. Set $x_\theta = (x_1 - x_0)/\|x_1 - x_0\|$. Then $\|x_\theta\| = 1$ and if $x \in X_0$, then $\|x_1 - x_0\| x + x_0 \in X_0$ so

$$\|x - x_\theta\| = \|x - x_1/\|x_1 - x_0\| + x_0/\|x_1 - x_0\|\|$$
$$= \tfrac{1}{\|x_1 - x_0\|} \|(\|x_1 - x_0\| x + x_0) - x_1\| \geq \tfrac{1}{\|x_1 - x_0\|} d$$

But, $d/\|x_1 - x_0\| \geq \theta$ implies $\|x - x_\theta\| \geq \theta$ for every $x \in X_0$.

Example 4.7. In general, x_θ cannot be chosen to be distance 1 from X_0 although this is the case when X_0 is finite dimensional. Let $X \subset C[0,1]$ be the closed subspace of those functions f such that $f(0) = 0$. Set $X_0 = \{f \in X : \int_0^1 f = 0\}$. Suppose there exists $f_1 \in X$ such that $\|f_1\| = 1$ and $\|f - f_1\| \geq 1$ for all $f \in X_0$. For $f \in X \setminus X_0$, let $c = \int_0^1 f_1 / \int_0^1 f$. Then $f_1 - cf \in X_0$ and, therefore, $1 \leq \|f_1 - (f_1 - cf)\| = |c| \|f\|$ which implies $\left|\int_0^1 f\right| \leq \|f\| \left|\int_0^1 f_1\right|$. Now we can make $\left|\int_0^1 f\right|$ as close to 1 as we please and still have $\|f\| = 1$ [$f_j(t) = t^{1/j}$ as $j \rightarrow \infty$ will work]. Thus, $1 \leq \left|\int_0^1 f_1\right|$. But, since $\|f_1\| = 1$ and $f_1(0) = 0$, $\left|\int_0^1 f_1\right| < 1$.

Theorem 4.8. *Let X be a normed space and suppose the unit ball $S = \{x :$ $\|x\| \le 1\}$ is compact. Then X is finite dimensional.*

Proof: Suppose that X is not finite dimensional and let $x_1 \in X$ with X_1 the subspace spanned by x_1. Then $X_1 \subsetneq X$ and X_1 is closed by Corollary 3. By Riesz's Lemma, there exists $x_2 \in S$ such that $\|x_2 - x_1\| \ge 1/2$. Let X_2 be the subspace spanned by $\{x_1, x_2\}$. Then $X_2 \subsetneq X$ and X_2 is closed. By Riesz's Lemma, there exists $x_3 \in X$ such that $\|x_3 - x\| \ge 1/2$ for every $x \in X_2$. In particular, $\|x_3 - x_j\| \ge 1/2$ for $j = 1, 2$. Inductively, there exists a sequence $\{x_j\} \subset X$ such that $\|x_j - x_k\| \ge 1/2$ for $j \ne k$. That is, $\{x_j\}$ has no convergent subsequences and, hence, S is not compact.

Corollary 4.9. *Let X be a normed linear space. Then X is finite dimensional iff the closed unit ball of X is compact.*

Exercises.

1. Show that a normed linear space is finite dimensional iff closed, bounded subsets are compact.

2. Let X, Y be normed linear spaces and $T : X \to Y$ be linear. Show that if X is finite dimensional, then T is continuous. Show that if T has a closed kernel and Y is finite dimensional, then T is continuous.

3. Are the norms $\|\cdot\|_\infty$ and $\|f\|_1 = \int_a^b |f|$ on $C[a, b]$ equivalent?

4. Let X be a vector space with 2 equivalent norms $\|\cdot\|_1, \|\cdot\|_2$. Show that the dual norms associated with them are also equivalent.

5. Let $\|\cdot\|_i$ be norms on a vector space X ($i = 1, 2, 3$). Show that if $\|\cdot\|_1, \|\cdot\|_2$ are equivalent and $\|\cdot\|_2, \|\cdot\|_3$ are equivalent, then $\|\cdot\|_1, \|\cdot\|_3$ are equivalent.

6. Show that a Banach space cannot have countable algebraic dimension. (Hint: Use the Baire Category Theorem (Appendix D).) Show completeness is important.

7. A series $\sum_j x_j$ in a normed space X is subseries convergent if for every subsequence $\{n_j\}$, the subseries $\sum_{j=1}^\infty x_{n_j}$ converges. Show that if X is finite dimensional, then every subseries convergent series is absolutely convergent. Give an example of a series which is subseries convergent but not absolutely convergent (Hint: Consider c_0.). An interesting result of Dvoretsky/Rogers states that a Banach space X is finite dimensional iff every subseries convergent series is absolutely convergent (see [LT]).

Chapter 5

Inner Product and Hilbert Spaces

In this chapter we study inner product and Hilbert spaces. This material will be used extensively in later chapters on continuous linear operators and spectral theory.

Let X be a vector space.

Definition 5.1. An inner product on X is a function $\cdot : X \times X \to \Bbbk$ such that

(i) $(x_1 + x_2) \cdot x_3 = x_1 \cdot x_3 + x_2 \cdot x_3$ for $x_i \in X$
(ii) $x_1 \cdot x_2 = \overline{x_2 \cdot x_1}$ for $x_i \in X$
(iii) $(tx_1) \cdot x_2 = t(x_1 \cdot x_2)$ for $x_i \in X, t \in \Bbbk$
(iv) $x \cdot x \geq 0$ for $x \in X$ and $x \cdot x = 0$ iff $x = 0$..

Note that (ii) and (iii) imply $x_1 \cdot (tx_2) = \bar{t}(x_1 \cdot x_2)$ and (i) and (ii) imply $x_1 \cdot (x_2 + x_3) = x_1 \cdot x_2 + x_1 \cdot x_3$.

A vector space X with an inner product is called an *inner product space*.

Theorem 5.2. *(Schwarz Inequality) Let X be an inner product space. Then $|x \cdot y| \leq \sqrt{x \cdot x}\sqrt{y \cdot y}$ for $x, y \in X$.*

Proof: If $y = 0$, the result is trivial so assume $y \neq 0$. In this case the inequality is equivalent to
$$|x \cdot y / \sqrt{y \cdot y}| \leq \sqrt{x \cdot x}$$
so we may assume $y \cdot y = 1$. Then
$$(*) \ 0 \leq (x - (x \cdot y)y) \cdot (x - (x \cdot y))y) = x \cdot x + |x \cdot y|^2 - (x \cdot y)(y \cdot x) - \overline{(x \cdot y)}(x \cdot y)$$
$$= x \cdot x + |x \cdot y|^2 - (x \cdot y)\overline{(x \cdot y)} - |x \cdot y|^2 = x \cdot x - |x \cdot y|^2.$$

Remark 5.3. Equality holds in the Schwarz inequality if x and y are linearly independent. The converse also holds for if equality holds in the Schwarz inequality with $y \neq 0$, then $(*)$ implies $x - (x \cdot y)y = 0$.

Theorem 5.4. *If X is an inner product space, then $x \to \sqrt{x \cdot x} = \|x\|$ defines a norm on X.*

Proof: Only the triangle inequality needs to be checked. For $x, y \in X$, we have

$$\|x + y\|^2 = (x + y) \cdot (x + y) = \|x\|^2 + \|y\|^2 + x \cdot y + y \cdot x$$
$$\leq \|x\|^2 + \|y\|^2 + 2 \|x\| \|y\| = (\|x\| + \|y\|)^2$$

by the Schwarz inequality.

If X is an inner product space we always assume that X is equipped with the norm from Theorem 4.

Proposition 5.5. *If X is an inner product space, then the inner product is a continuous function from $X \times X \to \Bbbk$.*

Proof: If $x_j \to x$ and $y_j \to y$, then

$$|x_j \cdot y_j - x \cdot y| \leq |x_j \cdot y_j - x_j \cdot y| + |x_j \cdot y - x \cdot y| \leq \|x_j\| \|y_j - y\| + \|y\| \|x_j - x\|$$

by the Schwarz inequality.

We have the following important property of the norm in an inner product space.

Proposition 5.6. *(Parallelogram Law) If X is an inner product space, then $\|x + y\|^2 + \|x - y\|^2 = 2 \|x\|^2 + 2 \|y\|^2$.*

Proof: $\|x + y\|^2 + \|x - y\|^2 = (x + y) \cdot (x + y) + (x - y) \cdot (x - y)$
$= 2 \|x\|^2 + 2 \|y\|^2 + x \cdot y + y \cdot x - x \cdot y - y \cdot x.$

Remark 5.7. The Parallelogram Law characterizes inner product spaces among the class of normed spaces. That is, if X is a normed space whose norm satisfies the Parallelogram Law, then the norm is induced by an inner product. If X is a vector space over $\mathbb{R}$, the inner product is defined by $4x \cdot y = \|x + y\|^2 - \|x - y\|^2$; if X is complex, the inner product is defined by $4x \cdot y = \|x + y\|^2 - \|x - y\|^2 + i \|x + iy\|^2 - i \|x - iy\|^2$. We leave the (tedious!) verifications to the reader.

Definition 5.8. An inner product space which is a Banach space under the induced norm is called a Hilbert space.

Example 5.9. $\mathbb{R}^n$ ($\mathbb{C}^n$) is a Hilbert space under the inner product

$$x \cdot y = (x_1, ..., x_n) \cdot (y_1, ..., y_n) = \sum_{j=1}^{n} x_j \overline{y_j}.$$

Example 5.10. l^2 is a Hilbert space under the inner product

$$s \cdot t = \{s_j\} \cdot \{t_j\} = \sum_{j=1}^{\infty} s_j \overline{t_j}.$$

This is the original Hilbert space.

Example 5.11. $C_{\mathbb{C}}[a, b]$ is an inner product space under the inner product $f \cdot g = \int_a^b f\overline{g}$. This space is not complete under the induced norm. This lack of completeness is one of the principal reasons that the Lebesgue integral is employed.

We establish an important geometric property of Hilbert space. A subset K of a vector space X is *convex* if $x, y \in K$ and $0 < t < 1$ implies $tx + (1 - t)y \in K$. That is, if $x, y \in K$, then the line segment from x to y also lies in K.

Theorem 5.12. *Let K be a non-empty, closed, convex subset of a Hilbert space H. If $x \in H$, then there is a unique $y \in K$ such that $\|x - y\| = \min\{\|x - z\| : z \in K\} = dist(x, K)$. Furthermore, y can be characterized by:*

$$(\#) \quad y \in K, \mathcal{R}(x - y) \cdot (z - y) \le 0 \text{ for all } z \in K.$$

Proof: Set $d = dist(x, K) \ge 0$. If $w, z \in K$ applying the parallelogram law to $(x - z)/2$ and $(x - w)/2$ gives

$$(*) \quad d^2 \le \|(w + z)/2 - x\|^2 = \|w - x\|^2/2 + \|z - x\|^2/2 - \|(w - z)/2\|^2$$

since $(w + z)/2 \in K$.

If $\|z - x\| = d$ and $\|w - x\| = d$, then $(*)$ implies $w = z$ so uniqueness holds.

Pick $\{y_k\} \subset K$ such that $\|x - y_k\| \to d$. Set $w = y_k, z = y_j$ in $(*)$ to obtain $\|(y_k - y_j)/2\| \le \|y_k - x\|^2/2 + \|y_j - x\|^2/2 \to 0$. Thus, $\{y_k\}$ is a Cauchy sequence in H and converges to some $y \in K$ with $\|x - y\| = d$.

If $y \in K$ satisfies $\|y - x\| = d$, then for $z \in K$ and $0 < t < 1$,

$$\|x - y\| \le \|x - tz - (1 - t)y\| = \|x - y - t(z - y)\|$$

so

$$\|x - y\|^2 \le \|x - y\|^2 + t^2 \|z - y\|^2 - t(x - y) \cdot (z - y) - t(z - y) \cdot (x - y)$$

or

$$2\mathcal{R}(x-y)\cdot(z-y) \le t\,\|z-y\|^2.$$

Letting $t \to 1$ gives (#).

On the other hand, if y satisfies (#) and $z \in K$, then

$$\|x-y\|^2 - \|x-z\|^2 = 2\mathcal{R}(x-y)\cdot(z-y) - \|z-y\|^2 \le 0$$

so $\|x-y\| = d$.

Let P_K be the "projection" map which sends x onto y in Theorem 12 above. If $H = \mathbb{R}^2$, inequality (#) means that the vector from x to $P_K x$ makes an obtuse angle with the vector from z to $P_K x$. Moreover, this map is uniformly continuous on H since $\|P_K u - P_K v\| \le \|u - v\|$ [from (#) $\mathcal{R}(u - P_K u)\cdot(P_K v - P_K u) \le 0$ and $\mathcal{R}(v - P_K v)\cdot(P_K u - P_K v) \le 0$ so adding gives $\mathcal{R}(u - v - (P_K u - P_K v))\cdot(P_K v - P_K u) \le 0$ so by the Schwarz inequality $\|P_K u - P_K v\|^2 \le \mathcal{R}(u - v)\cdot(P_K v - P_K u)$].

If X is an inner product space, then two elements x and y of X are *orthogonal* if $x \cdot y = 0$; we write $x \perp y$. A subset $S \subset X$ is *orthogonal* if $x \perp y$ for all $x, y \in S$. If S is orthogonal, then S is *orthonormal* if $\|x\| = 1$ for all $x \in S$.

Example 5.13. In l^2, $\{e^j : j \in \mathbb{N}\}$ is orthonormal.

Example 5.14. In $C_{\mathbb{C}}[-\pi, \pi]$, the space of complex valued continuous functions on $[-\pi, \pi]$, $\{e^{ikt}/\sqrt{2\pi} : k = 0, \pm 1, \pm 2, ...\}$ is orthonormal.

Example 5.15. In $C[0, \pi]$, $\{\sqrt{2}\sin k\pi t : k \in \mathbb{N}\}$ is orthonormal.

The Gram-Schmidt procedure (Exercise 5) can be applied to obtain important classes of orthonormal polynomials such as the Legendre, Hermite and Laguerre polynomials (see [DM] for explicit expressions for these polynomials). See also the orthonormal system described in Remark 30.

Proposition 5.16. *Let X be an inner product space. If $x \perp y$, then* $\|x + y\|^2 = \|x - y\|^2 = \|x\|^2 + \|y\|^2$.

Proof: $(x + y)\cdot(x + y) = \|x + y\|^2 = \|x\|^2 + \|y\|^2 = \|x - y\|^2$.

If X is an inner product space and $M \subset X$, then the *orthogonal complement* of M is defined to be

$$M^\perp = \{x \in X : x \perp y \text{ for all } y \in M\}.$$

Note that $M^\perp$ is a closed linear subspace. We now use Theorem 12 to show that any closed linear subspace of a Hilbert space has a complement.

Theorem 5.17. *If M is a closed linear subspace of a Hilbert space H, then $H = M \oplus M^\perp$.*

Proof: For $x \in H$, let $y = P_M x \in M$ be as in Theorem 12. Then $x - y \in M^\perp$ since by (#) $\mathcal{R}(x - y) \cdot w \leq 0$ for $w \in M$. Hence, $x = (x - y) + y$ with $x - y \in M^\perp$ and $y \in M$. Since $M \cap M^\perp = \{0\}$, we have $H = M \oplus M^\perp$.

Thus, every closed linear subspace of a Hilbert space has a closed complemented subspace. This actually characterizes Hilbert spaces among Banach spaces (this is non-trivial and was a long standing unsolved problem; see [LT]).

We now describe the dual of a Hilbert space.

Proposition 5.18. *Let H be a Hilbert space and $y \in H$. If $f_y : H \to \Bbbk$ is defined by $f_y(x) = x \cdot y$, then $f_y \in H'$ and $\|f_y\| = \|y\|$.*

Proof: f_y is clearly linear and since $|f_y(x)| = |x \cdot y| \leq \|x\| \, \|y\|$ it follows that $f_y \in H'$ with $\|f_y\| \leq \|y\|$. Since $f_y(y) = y \cdot y = \|y\|^2$, $\|f_y\| = \|y\|$.

Thus, the map $y \to f_y$ is an isometry from H into H'. We show that this map is onto.

Theorem 5.19. *(Riesz Representation Theorem) If H is a Hilbert space and $f \in H'$, then there exists a unique $y \in H$ such that $f = f_y$.*

Proof: If $f = 0$, put $y = 0$. Suppose $f \neq 0$. Set $M = \ker(f)$ so M is a proper closed subspace of H and $M^\perp \neq \{0\}$. Choose $z \in M^\perp$, $z \neq 0$. Set $y = (\overline{f(z)}/ \|z\|^2)z$ so $y \in M^\perp$, $y \neq 0$ and $f(y) = |f(z)|^2 / \|z\|^2$. For $x \in H$ let $x_1 = x - (f(x)/ \|y\|^2)y$ and $x_2 = (f(x)/ \|y\|^2)y$ so $x = x_1 + x_2$ and $f(x_1) = 0$ so $x_1 \in M$ and $x_1 \cdot y = 0$. Hence, $x \cdot y = x_2 \cdot y = f(x) = f_y(x)$ and $f = f_y$.

The map $\Theta_H = \Theta : H \to H'$ is an isometry from H onto H' which is additive but is only conjugate homogeneous in the sense that $\Theta(ty) = \bar{t}\Theta(y)$. Thus, if H is a real Hilbert space, then H and H' are linearly isometric and $H = H'$.

If X is an inner product space and $E = \{x_a : a \in A\}$ is an orthonormal subset of X, then for every $x \in X$ the scalars $\hat{x}(a) = x \cdot x_a, a \in A$, are called the *Fourier coefficients* of x with respect to E. We establish several important properties of the Fourier coefficients.

Proposition 5.20. *Let X be an inner product space and $\{x_1, ..., x_n\}$ an orthonormal subset of X. Then for every $x \in X$,*

(i) $\sum_{j=1}^{n} |x \cdot x_j|^2 = \sum_{j=1}^{n} \left|\hat{x}(j)\right|^2 \leq \|x\|^2$,

(ii) $(x - \sum_{j=1}^{n}(x \cdot x_j)x_j)\perp x_j$ *for every j.*

Proof: (i): $0\leq \left\|x - \sum_{j=1}^{n}(x \cdot x_j)x_j\right\|^2 =$
$(x - \sum_{j=1}^{n}(x \cdot x_j)x_j) \cdot (x - \sum_{j=1}^{n}(x \cdot x_j)x_j) =$
$\|x\|^2 - \sum_{j=1}^{n}(x \cdot x_j)\overline{(x \cdot x_j)} - \sum_{j=1}^{n}\overline{(x \cdot x_j)}(x \cdot x_j) + \sum_{j=1}^{n}\sum_{i=1}^{n}(x \cdot x_i)\overline{(x \cdot x_j)}x_i \cdot x_j$
$= \|x\|^2 - \sum_{j=1}^{n}|x \cdot x_j|^2$.
(ii): $(x - \sum_{j=1}^{n}(x \cdot x_j)x_j) \cdot x_i = x \cdot x_i - \sum_{j=1}^{n}(x \cdot x_j)(x_j \cdot x_i) = x \cdot x_i - x \cdot x_i = 0$.

We generalize the inequality in (i) to infinite sets.

Proposition 5.21. *Let $E = \{x_a : a \in A\}$ be an orthonormal subset of an inner product space X. For each $x \in X$ the set $E_x = \{a \in A : x \cdot x_a \neq 0\}$ is at most countable.*

Proof: For $n \in \mathbb{N}$ let $S_n = \{a \in A : |x \cdot x_a| > \|x\|^2/n\}$. By Proposition 20 S_n contains at most $n - 1$ elements. Since $E_x = \cup_{n=1}^{\infty}S_n$, the result follows.

Theorem 5.22. *(Bessel's Inequality) Let $E = \{x_a : a \in A\}$ be an orthonormal subset of an inner product space X. For each $x \in X$, $\sum_{a \in A}|x \cdot x_a|^2 \leq \|x\|^2$.*

Proof: If A is finite this is Proposition 20. If A is infinite, we must assign a meaning to $\sum_{a \in A}|x \cdot x_a|^2$. Let $S = \{a \in A : x \cdot x_a \neq 0\}$. If $S = \emptyset$, we set $\sum_{a \in A}|x \cdot x_a|^2 = 0$. If S is finite, we set $\sum_{a \in A}|x \cdot x_a|^2 = \sum_{a \in S}|x \cdot x_a|^2$ and the inequality follows from Proposition 20. If S is infinite, S is countable by Proposition 21 so the elements $\{x_a : a \in S\}$ can be arranged in a sequence $\{y_1, y_2, ...\}$. By Proposition 20 for every n, $\sum_{j=1}^{n}|x \cdot y_j|^2 \leq \|x\|^2$ so the series $\sum_{j=1}^{\infty}|x \cdot y_j|^2$ is absolutely convergent and its sum is independent of the ordering of the elements $\{x_a : a \in S\}$. Therefore, we may define

$\sum_{a \in A} |x \cdot x_a|^2 = \sum_{j=1}^{\infty} |x \cdot y_j|^2$ and $\sum_{a \in A} |x \cdot x_a|^2 \leq \|x\|^2$ by Proposition 20.

We will show that equality holds in Bessel's inequality for certain orthonormal subsets in a Hilbert space. An orthonormal subset E of a Hilbert space H is *complete* (or a complete orthonormal set) if $E_1 \subset H$ orthonormal and $E \subset E_1$ implies that $E = E_1$ (that is, E is maximal with respect to set inclusion). A complete orthonormal set is called an *orthonormal basis*. We give several criteria for an orthonormal subset to be complete. First, we need a lemma.

Lemma 5.23. *Let $\{x_1, ..., x_n\}$ be an orthonormal subset in an inner product space X.*

(i) *If $x = \sum_{j=1}^{n} c_j x_j$, then $c_j = x \cdot x_j = \hat{x}(j)$ and $\|x\|^2 = \sum_{j=1}^{n} |c_j|^2$.*

(ii) *For $\{c_1, ..., c_n\} \subset \Bbbk$, $\left\| x - \sum_{j=1}^{n} c_j x_j \right\|$ (as a function of $c_1, ..., c_n$) attains its minimum at $c_j = x \cdot x_j = \hat{x}_j$, $j = 1, ..., n$.*

Proof: (i): That $c_j = x \cdot x_j$ is clear.

$$\|x\|^2 = x \cdot x = \sum_{j=1}^{n} \sum_{k=1}^{n} c_k \overline{c_j} x_k \cdot x_j = \sum_{j=1}^{n} |c_j|^2 .$$

(ii):

$$0 \leq \left\| x - \sum_{j=1}^{n} c_j x_j \right\|^2 = \left(x - \sum_{j=1}^{n} c_j x_j \right) \cdot \left(x - \sum_{j=1}^{n} c_j x_j \right)$$
$$= \|x\|^2 - \sum_{j=1}^{n} c_j (x \cdot x_j) - \sum_{j=1}^{n} \overline{c_j} (x \cdot x_j) + \sum_{j=1}^{n} |c_j|^2$$
$$= \left(\|x\|^2 - \sum_{j=1}^{n} |x \cdot x_j|^2 \right) + \sum_{j=1}^{n} (x \cdot x_j - c_j)(\overline{x \cdot x_j - c_j})$$

and the expression on the right hand side of this equality is clearly minimal at $c_j = x \cdot x_j$.

Theorem 5.24. *Let $E = \{x_a : a \in A\}$ be an orthonormal subset of a Hilbert space H. The following are equivalent:*

(i) *E is complete.*

(ii) *$x \perp x_a$ for all $a \in A$ implies $x = 0$.*

(iii) *span$E$ is dense in H.*

(iv) *If $x \in H$, then $\|x\|^2 = \sum_{a \in A} |x \cdot x_a|^2$ (equality in Bessel's inequality).*

(v) $x = \sum_{a \in A} (x \cdot x_a) x_a$ for all $x \in H$.

(vi) If $x, y \in H$, then $x \cdot y = \sum_{a \in A} (x \cdot x_a)\overline{(y \cdot x_a)}$ (Parseval's Equality).

Proof: (i)$\Rightarrow$(ii): If (ii) is false, there exists $x \neq 0$ such that $x \perp x_a$ for all $a \in A$. Set $z = x/\|x\|$. Then $\{z\} \cup E$ is an orthonormal set properly containing E so E is not maximal and (i) fails.

(ii)$\Rightarrow$(iii): Let M be the closure of $spanE$. If $M \neq H$, then $H = M \oplus M^\perp$ with $M^\perp \neq \emptyset$. If $x \neq 0, x \in M^\perp$, then $x \perp x_a$ for all $a \in A$ so (ii) fails.

(iii)$\Rightarrow$(iv): Let $\epsilon > 0$ and $x \in H$. There exist $x_{a_1}, ..., x_{a_n} \in E$ and $c_1, ..., c_n$ such that $\left\| x - \sum_{j=1}^n c_j x_{a_j} \right\| < \epsilon$. By Lemma 23 (ii), $\left\| x - \sum_{j=1}^n (x \cdot x_j) x_{a_j} \right\| < \epsilon$. Thus, from Proposition 1.3, $(\|x\| - \epsilon)^2 \leq \left\| \sum_{j=1}^n (x \cdot x_j) x_{a_j} \right\|^2 = \sum_{j=1}^n |x \cdot x_{a_j}|^2 \leq \sum_{a \in A} |x \cdot x_a|^2$. Bessel's inequality gives the reverse inequality.

(iv)$\Rightarrow$(v): As in Theorem 22 let $S = \{x_a : x_a \cdot x \neq 0\}$ and arrange the elements of S into a sequence $y_1, y_2, ...$ Then

$$\left\| x - \sum_{j=1}^n (x \cdot y_j) y_j \right\|^2 = \|x\|^2 - \sum_{j=1}^n |x \cdot y_j|^2 =$$

$$\sum_{j=1}^\infty |x \cdot y_j|^2 - \sum_{j=1}^n |x \cdot y_j|^2 = \sum_{j=n+1}^\infty |x \cdot y_j|^2$$

by (iv). Hence, $x = \sum_{j=1}^\infty (x \cdot y_j) y_j = \sum_{a \in A} (x \cdot x_a) x_a$.

(v)$\Rightarrow$(vi): By Proposition 5, $x \cdot y = \sum_{a \in A} \sum_{b \in A} (x \cdot x_a)\overline{(y \cdot x_b)} x_a \cdot x_b = \sum_{a \in A} (x \cdot x_a)\overline{(y \cdot x_a)}$.

(vi)$\Rightarrow$(i): If (i) fails, there exists $z \in H$ with $\|z\| = 1$ and $z \perp x_a$ for all $a \in A$. Then $z \cdot z = 1$ while $\sum_{a \in A} |z \cdot x_a|^2 = 0$ so (vi) fails.

Example 5.25. $\{e^j : j \in \mathbb{N}\}$ is a complete orthonormal subset of l^2.

Theorem 5.26. *Let H be a Hilbert space. If E and F are complete orthonormal subsets of H, then E and F have the same cardinality.*

Proof: Since orthonormal subsets are linearly independent, we may assume that E and F are infinite.

For $e \in E$, let $F_e = \{f \in F : f \cdot e \neq 0\}$. By Theorem 24(ii) $F = \cup_{e \in E} F_e$ and by Proposition 21 each F_e is at most countable. Hence, the cardinality of F is less than or equal to the cardinality of E. Symmetry gives the reverse inequality.

The cardinality of a (any) complete orthonormal set is called the *orthonormal dimension* of the Hilbert space.

We next establish a result on the stability of orthonormal bases. For this we first require a lemma.

Lemma 5.27. *Let M, N be subspaces of an inner product space X with $\dim M < \dim N$. Then $M^{\perp} \cap N \neq \emptyset$.*

Proof: Let $x_1, ..., x_m$ be a basis for M and let $y_1, ..., y_{m+1}$ be linearly independent in N. It suffices to show there exist scalars $t_1, ..., t_{m+1}$ not all 0 such that $\sum_{i=1}^{m+1} t_i y_i \perp x_k$ for all $k = 1, ..., m$ or $\sum_{i=1}^{m+1} t_i y_i \cdot x_k = 0$ for $k = 1, ..., m$. But, this is a linear system of m equations in $m+1$ unknowns $t_1, ..., t_{m+1}$ and so has a non-trivial solution.

We now establish our stability result which states that if an orthonormal set is "close enough" to an orthonormal basis it is an orthonormal basis.

Theorem 5.28. *Let $\{x_1, ...\}$ be an orthonormal basis for a Hilbert space H and let $\{y_1, ...\}$ be an orthonormal set such that $\sum_{i=1}^{\infty} \|x_i - y_i\|^2 < \infty$. Then $\{y_i, ...\}$ is an orthonormal basis.*

Proof: Suppose $\{y_1, ...\}$ is not an orthonormal basis. By Theorem 24 there exists $y_0 \in H, y_0 \neq 0$ such that $y_0 \perp y_j$ for $j \geq 1$. Choose n such that $\sum_{j=n}^{\infty} \|x_j - y_j\|^2 < 1$. By the Lemma there exists $w \neq 0, w \in span\{y_0, y_1, ..., y_n\}$ such that $w \perp x_j$ for $j = 1, ..., n$. Then

$$0 < \|w\|^2 = \sum_{j=1}^{\infty} |w \cdot x_j|^2 = \sum_{j=n+1}^{\infty} |w \cdot x_j|^2 = \sum_{j=n+1}^{\infty} |w \cdot (x_j - y_j)|^2$$

$$\leq \|w\|^2 \sum_{j=n+1}^{\infty} \|x_j - y_j\|^2 < \|w\|^2,$$

a contradiction.

Remark 5.29. If $L^2(E)$ is the space of square integrable functions described in Remark1.22, there is a natural inner product defined on $L^2(E)$ by $f \cdot g = \int_E fg$ and $L^2(E)$ is a Hilbert space under the induced norm $\|f\|_2 = (\int_E |f|^2)^{1/2}$. There is no difficulty in extending $L^2(E)$ to complex valued functions and the inner product is then defined by $f \cdot g = \int_E f\bar{g}$.

The orthonormal set of functions $\{e^{ijt}/\sqrt{2\pi} : j = 0, \pm 1, ...\}$ is a complete orthonormal set in $L^2[-\pi, \pi]$ (see [Sw1]6.6.20,[HS]16.32). It is also the case that the Legendre, Hermite and Laguerre polynomials are complete in the Lebesgue space L^2, and these facts are usually established by using properties of the Lebesgue integral; see, for example, [HS].

Remark 5.30. There is another very important orthonormal system called the Haar system which we now describe. Let $h = \chi_{[0,1/2)} - \chi_{[1/2,1)}$ and set $h_{jk}(t) = 2^{j/2}h(2^j t - k)$ for $j, k \in \mathbb{Z}$. The collection $\{h_{jk} : j, k \in \mathbb{Z}\}$ is called the Haar system. It is easily seen that the Haar system is orthonormal and is actually a complete orthonormal system in $L^2(\mathbb{R})$. Note that the elements of the Haar system are generated by the single function h and translations and dilations of h. There are similar orthonormal systems generated by a single function called wavelets which have proven to be quite useful and important in both analysis and engineering applications. See the American Mathematical Monthly article by Strichartz [St] for a nice discussion of wavelets and [Wa] for a detailed exposition.

Exercises.

1. Show that any orthonormal subset of a Hilbert space is contained in a complete orthonormal subset. (Use Zorn's Lemma.)

2. Show a Hilbert space is separable iff its orthonormal dimension is countable.

3. If D is a dense subset of an inner product space and $x \perp D$, show $x = 0$.

4. If M is a linear subspace of a Hilbert space H and Y is a Banach space, show that any continuous linear operator $T : M \to Y$ has a continuous linear extension from H into Y. (The analogue of this result for normed spaces is false; there is no continuous linear extension of the identity operator on c_0 from l^∞ to c_0. See Exercise 11.5.)

5. (Gram-Schmidt) Let $x_1, ..., x_n$ be linearly independent in an inner product space. Set $y_1 = x_1, y_k = x_k - \sum_{j=1}^{k-1}(x_k \cdot y_j)y_j$ for $k > 1$ and $z_k = y_k / \|y_k\|$. Show $\{z_k\}$ is orthonormal and $span\{z_k\} = span\{x_k\}$. Apply the *Gram-Schmidt* procedure to compute a few terms of the orthonormal sequence generated by the polynomials $\{1, t, t^2, ...\}$; the resulting polynomials are the *Legendre polynomials*.

6. Let $A \neq \emptyset$. Define $l^2(A)$ to be all $f : A \to \Bbbk$ such that $\{a \in A : f(a) \neq 0\}$ is at most countable and $\sum_{a \in A} |f(a)|^2 < \infty$ (defined as in Theorem 22). If $f, g \in l^2(A)$ show $f \cdot g = \sum_{a \in A} f(a)\overline{g(a)}$ defines an inner product and

$l^2(A)$ is a Hilbert space. Let o_a be the function with value 1 at a and 0 otherwise. Show $\{o_a : a \in A\}$ is a complete orthonormal subset of $l^2(A)$. Use Theorem 26 to show that any Hilbert space is linearly isometric to some $l^2(A)$.

7. Show that the sup-norm on $C[0,1]$ does not satisfy the parallelogram law.

8. Let $S \subset X$, an inner product space. Show $S^\perp$ is a closed subspace, $S \subset S^{\perp\perp}$ and $S^{\perp\perp} = S^{\perp\perp\perp}$. If M is a closed linear subspace of a Hilbert space, show $M^{\perp\perp} = M$.

9. Show that the dual of a Hilbert space is a Hilbert space. (Hint: Use the isometry Θ_H from the Riesz Representation Theorem to transfer the inner product on H to H'.)

10. If J is any interval in $\mathbb{R}$, show there is an orthonormal subset in $C(J)$.

11. Let $g \in C[a,b]$ and define $G : C[a,b] \to C[a,b]$ by $Gf = gf$. Show G is continuous with respect to the norm induced by the inner product on $C[a,b]$.

12. Let X be an inner product space. Show $|x \cdot y| = \|x\|\,\|y\|$ iff $y = 0$ or $x = ty$ for some t.

13. Let X be an inner product space. Show $\|x+y\| = \|x\| + \|y\|$ iff $x = ty$ for some t.

14. Let M_1, M_2 be subspaces of an inner product space X. Show $M_1 \perp M_2$ iff $\|m_1 + m_2\|^2 = \|m_1\|^2 + \|m_2\|^2$ for all $m_i \in M_i$.

15. If $\{x_j : j \in \mathbb{N}\}$ is an orthonormal subset of a Hilbert space H, show that for each $x \in H$ $y = \sum_{j=1}^\infty (x \cdot x_j)x_j$ exists and $(x - y)\perp x_j$ for all j.

16. If $\{x_j : j \in \mathbb{N}\}$ is an orthonormal subset of a Hilbert space H, show $\sum_{j=1}^\infty x_j$ converges iff $\sum_{j=1}^\infty \|x_j\| < \infty$.

17. If X_1, X_2 are inner product (Hilbert) spaces, show $X_1 \oplus X_2$ is an inner product (Hilbert) space under a natural inner product.

Chapter 6

The Hahn-Banach Theorem

In this chapter we consider the first of the three basic principles of functional analysis, the Hahn-Banach Theorem. This theorem is concerned with the extension of linear functionals which are dominated by sublinear functionals. Despite the esoteric appearance of this theorem it has a surprising number of applications to various areas of analysis. We give an application to Banach limits at the end of this chapter and several applications to normed spaces in the following chapter.

Definition 6.1. Let X be a vector space. A function $p : X \to \mathbb{R}$ is sublinear if $p(x + y) \le p(x) + p(y)$ for $x, y \in X$ and $p(tx) = tp(x)$ for $x \in X$ and $t \ge 0$.

A semi-norm is sublinear but not conversely (see Exercise 1).

We first consider the case of real vector spaces.

Theorem 6.2. *(Hahn-Banach; real case) Let X be a real vector space and $p : X \to \mathbb{R}$ be a sublinear functional. Let M be a linear subspace of X. If $f : M \to \mathbb{R}$ is a linear functional such that $f(x) \le p(x)$ for all $x \in M$, then there exists a linear functional $F : X \to \mathbb{R}$ such that $F(x) = f(x)$ for all $x \in M$ and $F(x) \le p(x)$ for all $x \in X$.*

Proof: Let $\mathcal{E}$ be the class of all linear extensions g of f such that $g(x) \le p(x)$ for all $x \in \mathcal{D}(g)$, the domain of g, with $M \subset \mathcal{D}(g)$. Note $\mathcal{E} \ne \emptyset$ since $f \in \mathcal{E}$. Partial order $\mathcal{E}$ by $g < h$ iff h is a linear extension of g. If $\mathcal{C}$ is a chain in $\mathcal{E}$, then $\cup_{g \in \mathcal{C}} g$ is clearly an upper bound for $\mathcal{C}$ so by Zorn's Lemma $\mathcal{E}$ has a maximal element F. The result will follow if we show $\mathcal{D}(F) = X$.

Suppose there exist $x_1 \in X \backslash \mathcal{D}(F)$. Let M_1 be the linear subspace spanned by $\mathcal{D}(F)$ and x_1. Thus, if $y \in M_1$, then y has a unique representation $y = m + tx_1$, $m \in \mathcal{D}(F), t \in \mathbb{R}$. If $z \in \mathbb{R}$, then $F_1(y) = F_1(m + tx_1) = F(m) + tz$ defines a linear functional on M_1 which extends F. If we can show that it is possible to choose z such that $F_1(y) \le p(y)$ for all $y \in M_1$, this will show that $F_1 \in \mathcal{E}$ and will contradict the maximality of F.

In order to have

$$(*) \quad F_1(y) = F_1(m + tx_1) = F(m) + tz \le p(y) = p(m + tx_1)$$

we must have for $t > 0$, $z \le -\frac{1}{t}F(m) + \frac{1}{t}p(y) = -F(\frac{m}{t}) + p(\frac{m}{t} + x_1)$ or since $m/t \in \mathcal{D}(F)$, if z satisfies

$$(**) \quad z \le -F(m) + p(m + x_1)$$

for all $m \in \mathcal{D}(F)$, then (*) holds for all $t \ge 0$. For $t < 0$, $z \ge -\frac{1}{t}F(m) + \frac{1}{t}p(m + tx_1) = F(-m/t) - p(-m/t - x_1)$, or since $-m/t \in \mathcal{D}(F)$, if z satisfies

$$(***) \quad z \ge F(m) - p(m - x_1)$$

for all $m \in \mathcal{D}(F)$, then (*) holds. Thus, z must satisfy

$$F(m_1) - p(m_1 - x_1) \le z \le -F(m_2) + p(m_2 + x_1)$$

for all $m_1, m_2 \in \mathcal{D}(F)$, i.e., we must have

$$(****) \quad F(m_1) - p(m_1 - x_1) \le -F(m_2) + p(m_2 + x_1)$$

for all $m_1, m_2 \in \mathcal{D}(F)$. But, $F(m_1 + m_2) = F(m_1) + F(m_2) \le p(m_1 + m_2) \le p(m_1 - x_1) + p(m_2 + x_1)$ so (****) does hold.

To obtain a complex version of the Hahn-Banach Theorem we need a result which is interesting in its own right, due to Bohnenblust and Sobczyk, which shows how to write a complex linear functional in terms of its real part.

Lemma 6.3. *(Bohnenblust/Sobczyk) Let X be a vector space over $\mathbb{C}$ and suppose that $F = f + ig$ is a linear functional on X. Then for $x \in X$, $F(x) = f(x) - if(ix)$ and f is $\mathbb{R}$-linear. Conversely, if f is an $\mathbb{R}$-linear functional on X , then $F(x) = f(x) - if(ix)$ defines a $\mathbb{C}$-linear functional on X.*

Proof: f and g are clearly $\mathbb{R}$-linear. Now $F(ix) = iF'(x)$ implies $f(ix) + ig(ix) = if(x) - g(x)$ so $f(ix) = -g(x)$ and $F(x) = f(x) - if(ix)$.
The converse is easily checked (see Exercise 2).

Theorem 6.4. *(Hahn-Banach Theorem; complex case) Let X be a vector space and $p : X \to \mathbb{R}$ a semi-norm. Let M be a linear subspace of X and $f : M \to \Bbbk$ a linear functional. If $|f(x)| \leq p(x)$ for all $x \in M$, then there exists a linear functional $F : X \to \Bbbk$ extending f such that $|F(x)| \leq p(x)$ for all $x \in X$.*

Proof: Suppose X is a real vector space. Then $f(x) \leq |f(x)| \leq p(x)$ for all $x \in M$. Theorem 2 implies there exists a linear extension F of f satisfying $F(x) \leq p(x)$ for all $x \in X$. But then $F(-x) = -F(x) \leq p(-x) = p(x)$ so $|F(x)| \leq p(x)$.

Suppose X is a complex vector space. Then $\mathcal{R}f$, the real part of f, is real linear functional on M such that $|\mathcal{R}f(x)| \leq |f(x)| \leq p(x)$ for all $x \in M$. By the first part, there exists a real linear functional f_1 on X such that f_1 extends $\mathcal{R}f$ and satisfies $|f_1(x)| \leq p(x)$ for all $x \in X$. Set $F(x) = f_1(x) - if_1(ix)$. Then F is $\mathbb{C}$-linear and extends f by Lemma 3.

For $x \in X$, write $F(x) = |F(x)| e^{i\theta}$. Then $|F(x)| = e^{-i\theta} F(x) = F(e^{-i\theta}x) = \mathcal{R}F(e^{-i\theta}x) = f_1(e^{-i\theta}x) \leq p(e^{-i\theta}x) = p(x)$.

We now give an application of the Hahn-Banach Theorem to Banach limits. A Banach limit is an extension of the usual limit functional on the space c of convergent sequences to the space of bounded sequences l^∞ which still satisfies the basic properties of limits. The limit functional $\lim : c \to \mathbb{R}$ is defined by $\lim(\{t_j\}) = \lim_j t_j$. The linear functional $\lim$ is continuous on c with $|\lim t| \leq \|t\|_\infty$ for every $t \in c$ (Exercise 2.10). Thus, $\|\lim\| \leq 1$ and actually $\|\lim\| = 1$ since $\lim e = 1$ when e is the constant sequence $\{1, 1, ...\}$. By the Hahn-Banach Theorem , $\lim$ has a linear extension $L : l^\infty \to \mathbb{R}$ such that $|Lt| \leq \|t\|_\infty$ for all $t \in l^\infty$ so $\|L\| = 1$. We call such an extension an *extended limit*. We derive some of the basic properties of extended limits.

Proposition 6.5. *Let L be an extended limit. Then (i) if $t \geq 0$ in l^∞ (i.e., if $t = \{t_j\}$, then $t_j \geq 0$ for all j), $Lt \geq 0$ and (ii) for $t \in l^\infty$, $\liminf t \leq Lt \leq \limsup t$.*

Proof: (i): Suppose $t \in l^\infty$, $t \neq 0$ and $t \geq 0$. Let $\epsilon > 0$. Pick i such that $t_i + \epsilon > \|t\|_\infty$ and let e be the constant sequence $\{1, 1, ...\}$. Then

$$L(t - (t_i + \epsilon)e) \geq - \|t - (t_i + \epsilon)e\|$$

so

$$L(t) - (t_i + \epsilon) \geq -\sup_j(t_i + \epsilon - t_j) = -((t_i + \epsilon) - \inf_j t_j)$$

which implies $L(t) \geq \inf t_j \geq 0$.

(ii): Let $t \in l^\infty$. Choose n such that $\inf_j t_j \leq t_n < \inf_j t_j + \epsilon$ so $t_j + \epsilon - t_n > 0$ for every j which implies by (i) that $L(t) + \epsilon - t_n \geq 0$ or $L(t) \geq \inf_j t_j - \epsilon$ so $L(t) \geq \inf_j t_j$. Similarly, $L(t) \leq \sup_j t_j$. These inequalities give (ii).

Let τ be the left shift operator defined on l^∞; $\tau(\{t_1, t_2, ...\}) = \{t_2, t_3, ...\}$. An important property of the limit functional on c is that $\lim(\tau(t)) = \lim t$. If an extended limit L satisfies this property on l^∞, it is called a *Banach limit*. We now show that such Banach limits exist. Recall that e is the constant sequence $\{1, 1, ...\}$ and bs is the sequence space of all bounded series (Example 1.9).

Lemma 6.6. *If $L : l^\infty \to \mathbb{R}$ is continuous, linear and satisfies (i) $\|L\| = 1$, (ii) $L(e) = 1$, (iii) $L(bs) = \{0\}$, then L is a Banach limit.*

Proof: Since $c_{00} \subset bs$, $L(c_{00}) = \{0\}$ and by continuity $L(c_0) = \{0\}$. Therefore, L is an extended limit [if $t \in c$, let $t_0 = \lim t_j$; then $t - t_0 e \in c_0$ so $L(t - t_0 e) = L(t) - t_0 L(e) = L(t) - t_0 = 0$]. If $t \in^\infty$, $t - \tau t \in bs$ so $L(t - \tau t) = 0$ or $L(t) = L(\tau t)$.

Let C be the Cesaro matrix $C = [c_{ij}]$, where $c_{ij} = 1/i$ for $1 \leq j \leq i$ and $c_{ij} = 0$ for $i < j$ and recall that $C : l^\infty \to l^\infty$, $\|C\| = 1$ and C preserves limits for sequences in c and $C(bs) \subset c_0$ [Exercise 2.6].

Theorem 6.7. *Banach limits exist.*

Proof: Let L be an extended limit. By Lemma 6, LC is a Banach limit.

This elegant proof of the existence of Banach limits is due to Bennett and Kalton ([BeK]). The original proof of Banach uses the Hahn-Banach Theorem and the construction of a special sublinear functional ([B]).

Finally, we consider a classic problem on moments. Suppose we are given a sequence $c_0, c_1, ...$ of real numbers. When does there exist a function of bounded variation g such that $c_n = \int_0^1 t^n dg(t)$, $n = 0, 1, ...$; i.e., in probability terms when do the moments of a distribution determine the distribution? Since the dual of $C[0, 1]$ is the space of (normalized) functions of bounded variation (Appendix C), this suggests an abstract formulation of the moment problem. Given a normed space X, a family of scalars $\{c_a : a \in A\}$ and a subset $\{x_a : a \in A\} \subset X$, when does there exist $x' \in X'$ such that $x'(x_a) = c_a$ for all $a \in A$? We use the Hahn-Banach Theorem to give an answer to this question.

Theorem 6.8. *There exists $x' \in X'$ such that $x'(x_a) = c_a$ for all $a \in A$ iff there exists $M > 0$ such that $\left|\sum_{a \in \sigma} t_a c_a\right| \leq M \left\|\sum_{a \in \sigma} t_a x_a\right\|$ for all finite subsets $\sigma \subset A$ and scalars t_a.*

Proof: $\Rightarrow$: For a finite subset $\sigma \subset A$, $\left|\sum_{a \in \sigma} t_a c_a\right| = \left|\sum_{a \in \sigma} t_a x'(x_a)\right| \leq \|x'\| \left\|\sum_{a \in \sigma} t_a x_a\right\|$.

$\Leftarrow$: Let X_0 be the span of $\{x_a : a \in A\}$. If $x = \sum_{a \in \sigma} t_a x_a$, $\sigma \subset A$ finite, belongs to X_0, define $x' : X_0 \to \Bbbk$ by $x'(x) = \sum_{a \in \sigma} t_a c_a$. Note that x' is well defined since if $x = \sum_{a \in \sigma} t_a x_a = \sum_{b \in \tau} s_b x_b$ and we set $r_i = t_i - s_i$ for $i \in \sigma \cap \tau$, $r_i = t_i$ for $i \in \sigma \backslash \tau$ and $r_i = -s_i$ for $i \in \tau \backslash \sigma$, then

$$\left|\sum_{a \in \sigma} t_a c_a - \sum_{b \in \tau} s_b c_b\right| = \left|\sum_{i \in \sigma \cup \tau} r_i c_i\right|$$

$$\leq M \left\|\sum_{i \in \sigma \cup \tau} r_i x_i\right\| = M \left\|\sum_{a \in \sigma} t_a x_a - \sum_{b \in \tau} s_b x_b\right\| = 0.$$

Since $\left|x'(\sum_{a \in \sigma} t_a x_a)\right| = \left|\sum_{a \in \sigma} t_a c_a\right| \leq M \left\|\sum_{a \in \sigma} t_a x_a\right\|$, x' is a continuous linear functional on X_0. Now extend x' to X by Theorem 4 so $|x'(x)| \leq M \|x\|$ for all $x \in X$ and x' is continuous.

It is interesting to note that a moment problem of F. Riesz motivated an early version of the Hahn-Banach Theorem due to E. Helly. Although Helly proved this version in $C[a, b]$, his proof is modern in spirit and is essentially the proof given above in Theorem 2. For an interesting discussion of Helly's paper which also contains a version of the Uniform Boundedness Theorem, see [Ho]. See [BK],[Di] for a discussion of the history of the Hahn-Banach Theorem.

In the next chapter, we give a number of applications of the Hahn-Banach Theorem to topics in normed spaces. There are a number of applications of the Hahn-Banach Theorem to other ares of analysis; see, for example, Taylor/Lay ([TL])III.3,11 or Conway ([Co]).

There is also a separation version of the Hahn-Banach Theorem. If X is a vector space and M and N are subsets of X, a linear functional f is said to separate M and N if there exists $c \in \mathbb{R}$ such that $\mathcal{R}f(M) \geq c \geq$

$\mathcal{R}f(N)$. One form of the basic separation theorem for normed spaces is the statement that if X is a normed space and M, N are disjoint convex subsets with M open, then there exists a continuous linear functional $f \in X'$ which separates M and N. If f is a non-zero linear functional on a vector space X, then any level set $\{x : f(x) = c\} = H_c$ is called a hyperplane. In the statement above if X is a real normed space, the set N lies in the half-space $\{x : f(x) \leq c\}$ while M lies in the half-space $\{x : f(x) \geq c\}$, i.e. the hyperplane H_c separates M and N. See [DS]V.1.12 for more precise statements and proofs. The separation version of the Hahn-Banach Theorem has numerous applications to optimization theory; see, for example, Luenberger ([Lu]).

Exercises.

1. Give an example of a sublinear functional which is not a semi-norm. (Hint: consider $\limsup$ on l^∞.)

2. Prove the converse in Lemma 3.

Chapter 7

Applications of the Hahn-Banach Theorem to Normed Spaces

In this chapter we give applications of the Hahn-Banach Theorem to normed spaces. We first have a direct consequence of the Hahn-Banach Theorem.

Theorem 7.1. *Let M be a linear subspace of a semi-normed space X. If $f \in M'$, then there exists $F \in X'$ such that F extends f and $\|f\| = \|F\|$.*

Proof: Define a semi-norm p on X by $p(x) = \|x\| \|f\|$. Then $|f(m)| \leq p(m)$ for $m \in M$. By the Hahn-Banach Theorem 6.4 there exists a linear functional F on X which extends f and which satisfies $|F(x)| \leq p(x) = \|x\| \|f\|$. Hence, $F \in X'$ and $\|F\| \leq \|f\|$. But, obviously, $\|f\| \leq \|F\|$.

The analogue of Theorem 1 is false for continuous linear operators between normed spaces. For example, the identity operator on c_0 has no continuous linear extension to l^∞ (this will be established in Chapter 11, Exercise 5).

Next, we establish a geometric property of normed spaces.

Theorem 7.2. *Let M be a subspace of a normed space X. Suppose there exists $x \in X$ such that $d = dist(x, M) > 0$. Then there exists $f \in X'$ such that $\|f\| = 1$, $f(M) = \{0\}$ and $f(x) = d$.*

Proof: Let M_1 be the subspace spanned by M and x. Define a linear functional f on M_1 by $f(m_1) = f(tx + m) = td$. Then $f(m) = 0$ for all $m \in M$ and $f(x) = d$. Now $f \in M_1'$ and $\|f\| = 1$. For if $t \neq 0$ and $m \in M$, then $\|tx + m\| = |t| \|x + m/t\| \geq |t| d = |f(tx + m)|$ so $f \in M_1'$ and $\|f\| \leq 1$. There exists $\{m_j\} \subset M$ such that $\|x - m_j\| \downarrow d$ so $d = f(x - m_j) \leq \|f\| \|x - m_j\| \downarrow \|f\| d$ which implies $\|f\| \geq 1$. Hence, $\|f\| = 1$.

Now use Theorem 1 to extend f to X.

Remark 7.3. Note Theorem 2 applies if M is closed and $x \notin M$.

Corollary 7.4. *Let X be a normed space and $x \in X, x \neq 0$. Then there exists $f \in X'$ such that $\|f\| = 1$ and $f(x) = \|x\|$. In particular, if $x, y \in X$ and $x \neq y$, then there exists $f \in X'$ such that $f(x) \neq f(y)$; that is, X' separates the points of X.*

Proof: Set $M = \{0\}$ in Theorem 2.

Corollary 7.5. *For every x belonging to a normed space X, $\|x\| = \max\{|f(x)| : f \in X', \|f\| = 1\}$.*

Proof: For every $f \in X'$ with $\|f\| = 1$, we have $|f(x)| \leq \|x\|$. By Corollary 4, there exists $f \in X'$, $\|f\| = 1$ such that $f(x) = \|x\|$.

Remark 7.6. Note that the formula for $\|x\|$ in Corollary 5 is dual to the definition of the dual norm on X'.

We next have an interesting imbedding theorem for normed spaces.

Theorem 7.7. *Let X be a normed space. Then there exists $S \neq \emptyset$ such that X is isometrically isomorphic to a subspace of $B(S)$.*

Proof: Let S be the closed unit ball of X', $\{f \in X' : \|f\| \leq 1\}$. For $x \in X$ define $Ux : S \to \Bbbk$ by $Ux(f) = f(x)$. By Corollary 5, $\|Ux\|_{\infty} = \sup\{|f(x)| : f \in S\} = \|x\|$ so U is an isometry and is clearly linear.

We next address the converse of Theorem 2.3.

Theorem 7.8. *Let $X \neq \{0\}$ and Y be normed spaces. Then $L(X, Y)$ is a Banach space iff Y is a Banach space.*

Proof: If Y is complete, then $L(X, Y)$ is complete by Theorem 2.3.

Assume $L(X, Y)$ is complete and let $\{y_j\}$ be a Cauchy sequence in Y. Choose $x_0 \in X$, $\|x_0\| = 1$. By Corollary 5, there exists $f \in X'$ such that $f(x_0) = 1 = \|x_0\|$. Define $T_j \in L(X, Y)$ by $T_j x = f(x) y_j$. Now

$$\|(T_j - T_k)x\| = \|f(x)(y_j - y_k)\| \leq \|f\| \|x\| \|y_j - y_k\|$$

for every $x \in X$ so $\|T_j - T_k\| \leq \|f\| \|y_j - y_k\|$. Therefore, $\{T_j\}$ is Cauchy in $L(X, Y)$. By hypothesis, there exists $T \in L(X, Y)$ such that $\|T_j - T\| \to 0$. But, $\|y_j - Tx_0\| = \|T_j x_0 - Tx_0\| \leq \|T_j - T\| \|x_0\|$ so $y_j \to Tx_0$.

Let X be a normed space. The dual of X' with the dual norm, $(X')'$, is denoted by X'' and is called the *bidual* of X. We always assume that the bidual X'' is equipped with the dual norm from X'. We now observe that

X can be isometrically imbedded in the bidual X''. For $x \in X$, define a linear functional $\hat{x}$ on X' by $\hat{x}(x') = x'(x)$, $x' \in X'$. By Corollary 5, $\hat{x} \in X''$ and $\left\|\hat{x}\right\| = \sup\{\left|\hat{x}(x')\right| : \|x'\| = 1\} = \|x\|$ so the linear map $J_X = J : x \to \hat{x}$ is an isometry from X into X''; this map is called the *canonical imbedding* of X into X''. It is common practice to identify X and JX under this imbedding.

Definition 7.9. A normed space X is reflexive if the canonical imbedding J from X into X'' is onto.

We will give examples of reflexive and non-reflexive spaces below. Note that any reflexive normed space must be complete (a Banach space) since dual spaces are always complete.

Example 7.10. The spaces l^p for $1 < p < \infty$ are reflexive by the descriptions of the duals given in Theorem 2.21.

Remark 7.11. Note that X and X'' must be isomorphically isometric via the canonical imbedding to be reflexive. This is the case for the spaces l^p $(1 < p < \infty)$ in Example 10 above. However, James has given an example of a Banach space X which is not reflexive but is isometrically isomorphic to its bidual ([J]).

Proposition 7.12. *Let X be a reflexive space. Then every continuous linear functional f on X attains its maximum on the closed unit ball of X.*

Proof: By Corollary 5 there exists $x'' \in X''$ such that $\|x''\| = 1$ and $x''(f) = \|f\|$. But, there exists $x \in X$ such that $x'' = Jx$.

Example 7.13. c_0 is not reflexive. Define $h : c_0 \to \mathbb{R}$ by $h(t) = h(\{t_j\}) = \sum_{j=1}^{\infty} t_j/j!$. Then $h \in c_0'$ and $\|h\| = \sum_{j=1}^{\infty} 1/j!$. However, there is no $\{t_j\} \in c_0$ such that $\|h\| = h(\{t_j\})$. Hence, c_0 is not reflexive by Proposition 12.

Theorem 7.14. *If the dual of a normed space X is separable, then X is separable.*

Proof: Let $\{f_j\}$ be dense in X'. For each j choose $x_j \in X$, $\|x_j\| = 1$, such that $|f_j(x_j)| \geq \|f_j\|/2$. The subspace X_1 spanned by $\{x_j : j \in \mathbb{N}\}$ is dense in X. For if this is not the case, by Theorem 2 there exists $f \in X'$ such that $\|f\| = 1$ and $f(X_1) = \{0\}$. Since $\{f_j\}$ is dense in X', there is a subsequence $\{f_{n_j}\}$ such that $f_{n_j} \to f$ and since

$$\left\|f_{n_j} - f\right\| \geq \left|(f_{n_j} - f)(x_{n_j})\right| = \left|f_{n_j}(x_{n_j})\right| \geq \left\|f_{n_j}\right\|/2,$$

$\|f_{n_j}\| \to 0$. But, $\|f_{n_j}\| \to \|f\| = 1$. This contradiction shows that X_1 is dense in X.

Example 7.15. The converse is false. Consider $X = l^1$ and $X' = l^\infty$ (Exercise 1.14).

Corollary 7.16. *Let X be reflexive. Then X is separable iff X' is separable.*

Example 7.17. l^1 and l^∞ are not reflexive.

A normed space X has a *completion* if there exists a Banach space Y and a linear isometry $U : X \to Y$ such that X is a dense subspace of Y. It is customary to identify X as a subspace of Y.

Theorem 7.18. *Every normed space X is a dense subspace of a Banach space $\hat{X}$, i.e., every normed space has a completion.*

Proof: Set $\hat{X}$ equal to the closure of JX in X'' (here we are identifying X and JX).

Theorem 7.19. *A Banach space X is reflexive iff X' is reflexive.*

Proof: Suppose X is reflexive. Let $x''' \in X'''$. Define $x' \in X'$ by $x' = x'''J_X$. For any $J_X x \in X''$, $x'''(J_X x) = x'(x) = J_X x(x') = J_{X'}x'(J_X x)$ so $x''' = J_{X'}x'$.

Suppose X' is reflexive. By Remark 3 if we show $x'''(J_X X) = \{0\}$ implies $x''' = 0$, then $J_X X$ is dense in X'' and must equal X'' since $J_X X$ is complete and closed. Let $x''' \in X'''$ and let $x' \in X'$ be such that $J_{X'}x' = x'''$. If $x'''(J_{X'}x') = J_{X'}x'(J_X x) = x'(x) = 0$ for all $x \in X$, then $x' = 0$ or $x''' = 0$.

Theorem 7.20. *A closed subspace M of reflexive space X is reflexive.*

Proof: Let $m'' \in M''$. Define $x'' \in X''$ by $x''(x') = m''(x'_M)$, where x'_M is the restriction of x' to M. Set $m = J_X^{-1}x''$. We show $m \in M$ and $J_M m = m''$.

First, suppose $m \in X \backslash M$. Then by Theorem 2 there exists $x' \in X'$ such that $x'(m) \neq 0$ and $x'(M) = \{0\}$. Thus, $x'_M = 0$ and $0 \neq x'(m) = x'(J_X^{-1}x'') = x''(x') = m''(x'_M) = m''(0) = 0$ a contradiction. Thus, $m \in M$.

For each $m' \in M'$ let m'_X be an extension of m' to X (Theorem 1). Then $m''(m') = x''(m'_X) = m'_X(J_X^{-1}x'') = m'_X(m) = J_M m(m')$ so $m'' = J_M m$.

Exercises.

1. Show a Hilbert space H is reflexive. (Hint: H' and H'' are Hilbert spaces by Exercise 5.9. Show the conjugate linear isometries Ψ_H and $\Psi_{H'}$ satisfy $J_H = \Psi_{H'}\Psi_H$.)

2. Show that if X is reflexive, then X' has no proper closed subspaces which separate the points of X.

3. If X is reflexive and X' contains a countable set which separates the points of X, show X' is separable.

4. Show $C[0,1]$ is not reflexive by showing $C[0,1]'$ is not separable. [Hint: Consider the evaluation functionals $\delta_t(f) = f(t), 0 \leq t \leq 1$.]

5. Suppose $A \subset X$. Show x belongs to the closure of the span of A iff every $f \in X'$ such that $f(A) = \{0\}$ satisfies $f(x) = 0$. (Hint: Use Theorem 2.)

6. Show every inner product space has a completion which is a Hilbert space.

Chapter 8

The Uniform Boundedness Principle

In this chapter we consider the second of the three fundamental principles of functional analysis, the Uniform Boundedness Principle. First, we establish a lemma.

Lemma 8.1. *Let Y be a semi-normed space, $m_{ij} \in Y$, and $\epsilon_{ij} > 0$ for $i, j \in \mathbb{N}$. Suppose the rows and columns of the infinite matrix $M = [m_{ij}]$ converge to 0. Then there exists a subsequence $\{n_j\}$ such that $\|m_{n_i n_j}\| < \epsilon_{ij}$ for $i \neq j$.*

Proof: Set $n_1 = 1$. There exists $n_2 > n_1$ such that $\|m_{kn_1}\| < \epsilon_{21}$ and $\|m_{n_1 k}\| < \epsilon_{12}$ for $k \geq n_2$. There exists $n_3 > n_2$ such that $\|m_{n_1 k}\| < \epsilon_{13}, \|m_{n_2 k}\| < \epsilon_{23}, \|m_{kn_1}\| < \epsilon_{31}, \|m_{kn_2}\| < \epsilon_{32}$ for $k \geq n_3$. Now just continue.

Theorem 8.2. *(Uniform Boundedness Principle) Let X be a Banach space and Y a normed space with $\Gamma \subset L(X, Y)$. If $\{Tx : T \in \Gamma\}$ is bounded in Y for every $x \in X$, then $\{\|T\| : T \in \Gamma\}$ is bounded (i.e., if Γ is pointwise bounded on X, then Γ is uniformly bounded on bounded subsets of X).*

Proof: If the conclusion fails, there exists a sequence $\{T_j\} \subset \Gamma$ such that $\|T_j\| > j^4$. Pick $x_j \in X$ such that $\|x_j\| \leq 1$ and $\|T_j x_j\| > j^4$. Consider the matrix

$$M = [m_{ij}] = [\frac{1}{i} T_i(\frac{x_j}{j^2})].$$

By the continuity of the T_i and the pointwise boundedness assumption, the rows and columns of M converge to 0. By Lemma 1 (with $\epsilon_{ij} = 1/2^{i+j}$) there is a subsequence $\{n_j\}$ satisfying the conclusion of the lemma; to avoid messy subscript notation later, assume that $n_j = j$ so that $\|m_{ij}\| < 1/2^{i+j}$ when $i \neq j$. Let $x = \sum_{j=1}^{\infty} x_j / j^2$; note that this series converges in X since it is absolutely convergent and X is complete (Theorem 1.21). Then we have

$$\left\|\frac{1}{i}T_i x\right\| = \left\|\sum_{j=1}^{\infty}\frac{1}{i}T_i(\frac{x_j}{j^2})\right\|$$

$$\geq \left\|\frac{1}{i}T_i(x_i/i^2)\right\| - \left\|\sum_{j=1,j\neq i}^{\infty}\frac{1}{i}T_i(\frac{x_j}{j^2})\right\| \geq i - \sum_{j=1}^{\infty}1/2^{i+j} = i - 1/2^i$$

which implies that $\{T_i x\}$ is not bounded.

The completeness in Theorem 2 is important.

Example 8.3. Define $f_j : c_{00} \to \mathbb{R}$ by $f_j(t) = \sum_{i=1}^{j} t_i$. Then $f_j \in c'_{oo}$, $\{f_j(t) : j \in \mathbb{N}\}$ is bounded for each $t \in c_{00}$ but $\|f_j\| = j$.

Corollary 8.4. *(Banach-Steinhaus) Let X be a Banach space and Y a normed space. If $\{T_j\} \subset L(X,Y)$ is such that $\lim_j T_j x = Tx$ exists for every $x \in X$, then $T \in L(X,Y)$ and $\|T\| \leq \liminf_j \|T_j\|$.*

Proof: For every $x \in X$, $\{T_j x\}$ is bounded so by Theorem 2, $\{\|T_j\|\}$ is bounded by, say, M. Thus, $\|T_j x\| \leq M\|x\|$ for $x \in X, j \in \mathbb{N}$. Hence, $\|Tx\| \leq M\|x\|$ for $x \in X$ which implies T is continuous. Also,

$$\|Tx\| = \lim_j \|T_j x\| \leq \liminf_j \|T_j\|\,\|x\|$$

so $\|T\| \leq \liminf_j \|T_j\|$.

Completeness is also important in Corollary 4.

Example 8.5. Define $f_j : c_{00} \to \mathbb{R}$ by $f_j(t) = f_j(t) = \sum_{i=1}^{j} t_i$. Then $f_j \in c'_{oo}$ and $\lim_j f_j(t) = \sum_{j=1}^{\infty} t_j = f(t)$ but f is not continuous.

An important result following from Theorem 2 is a useful criterion for boundedness in a normed space.

Theorem 8.6. *A subset A of a normed space X is bounded iff the set $\{f(x) : x \in A\}$ is bounded for each $f \in X'$.*

Proof: If A is bounded, let $\|x\| \leq M$ for each $x \in A$. If $f \in X'$, then $|f(x)| \leq M\|f\|$ for $x \in A$.

For the converse, let J be the canonical imbedding of X into X''. Since $\{f(x) : x \in A\}$ is bounded for each $f \in X'$, $\{Jx(f) : x \in A\}$ is bounded for each $f \in X'$. Since X' is complete, $\{\|Jx\| : x \in A\}$ is bounded by Theorem 2. That is, $\{\|x\| : x \in A\}$ is bounded.

We now give several applications of the Uniform Boundedness Principle.

Proposition 8.7. *Let* $1 \leq p < \infty$ *and* $\frac{1}{p} + \frac{1}{q} = 1$. *Suppose the series* $\sum_{j=1}^{\infty} s_j t_j$ *converges for every* $t = \{t_j\} \in l^p$. *Then* $s = \{s_j\} \in l^q$.

Proof: For each n define $f_n : l^p \to \mathbb{R}$ by $f_n(t) = \sum_{j=1}^{n} s_j t_j$. Then $f_n \in (l^p)' = l^q$ and $\lim_n f_n(t) = f(t) = \sum_{j=1}^{\infty} s_j t_j$. By Corollary 4, f is continuous so $f = \{s_j\} \in l^q$ (Theorems 2.20 and 2.21).

We use Proposition 7 and the Uniform Boundedness Principle to establish an automatic continuity result for matrix mappings between l^p spaces. Recall that if $A = [a_{ij}]$ is an infinite matrix and E, F are sequence spaces, the matrix A maps E into F, $A : E \to F$, if for each $t = \{t_j\} \in E$ the series $\sum_{j=1}^{\infty} a_{ij} t_j$ converges and $At = \{\sum_{j=1}^{\infty} a_{ij} t_j\} \in F$. We show that any matrix A which maps l^p $(1 \leq p < \infty)$ into l^q $(1 \leq q < \infty)$ is continuous. Such theorems are called automatic continuity results since there are no topological assumptions on the mappings. Other such automatic continuity results are established in 12.12 and 19.3.

Corollary 8.8. *Let* $1 \leq p < \infty$ *and* $1 \leq q < \infty$. *Suppose the infinite matrix* $A = [a_{ij}] : l^p \to l^q$. *Then* A *is continuous.*

Proof: Let a^i be the ith row of A. Since $\sum_{j=1}^{\infty} a_j^i t_j$ converges for every $t \in l^p$, a^i is continuous by Proposition 7. Therefore, $A^n : l^p \to l^q$ defined by $A^n t = \sum_{j=1}^{n} (a^j(t)) e^j$ is continuous. Since $A^n t \to At$ in l^q for $t \in l^p$, A is continuous by Corollary 4.

We next use the Uniform Boundedness Theorem to derive an important result in summability theory. One of the important problems in summability theory is to extend the notion of limit to divergent sequences. One method of accomplishing this is to employ infinite matrices. For example, if a sequence $\{t_j\}$ converges to t, then the sequence of arithmetic averages $\{\sum_{j=1}^{n} t_j / n\}$ also converges to t; however, the arithmetic averages of the divergent sequence $\{(-1)^j\}$ is transformed into a convergent sequence. This can be described by using the Cesaro matrix $C = [c_{ij}]$, $c_{ij} = 1/i$ for $1 \leq i \leq j$ and $c_{ij} = 0$ otherwise since $Ct = \{\sum_{j=1}^{n} t_j / n\}$. Then C maps c into c and preserves limits (Exercise 2.6) so C extends the notion of limit to some divergent sequences. We give a characterization of such matrices. An infinite matrix $A = [a_{ij}]$ is *conservative* if A maps c into c, i.e., if A maps convergent sequences into convergent sequences; A is *regular* if A is conservative and preserves limits. For example, the Cesaro matrix C above is regular. The Silvermann-Toeplitz Theorem characterizes regular matrices.

Theorem 8.9. *The matrix* $A = [a_{ij}]$ *is regular iff (i)* $\sup_i \sum_{j=1}^{\infty} |a_{ij}| < \infty$, *(ii)* $\lim_i a_{ij} = 0$ *for each* j *and (iii)* $\lim_i \sum_{j=1}^{\infty} a_{ij} = 1$.

Proof: If A is regular, (ii) and (iii) follow by setting $t = e^j$ and $t = (1, 1, ...)$. For (i) note that each row $\{a_{ij}\}_j$ induces a continuous linear functional $r^i : c \to \mathbb{R}$ with $\|r^i\| = \sum_{j=1}^{\infty} |a_{ij}|$ (Exercise 1). Since $A : c \to c$, $\lim_i r^i(t)$ exists for every $t \in c$. By Theorem 2, $\sup_i \|r^i\| = \sup_i \sum_{j=1}^{\infty} |a_{ij}| < \infty$.

Suppose (i), (ii) and (iii) hold and let $\epsilon > 0$. Then (i) implies that $\sum_{j=1}^{\infty} a_{ij}t_j$ converges for every $t \in c$. Let $t \in c$ with $\lim t = l$. Set

$$M = \max\{\sup_i \sum_{j=1}^{\infty} |a_{ij}|, \sup_j |t_j - l|\}.$$

Choose N such that $|t_j - l| < \epsilon$ for $j \geq N$. Then

$$(*) \quad \left|\sum_{j=1}^{\infty} a_{ij}t_j - l\right| \leq \sum_{j=1}^{N} |a_{ij}| \, |t_j - l| + \sum_{j=N+1}^{\infty} |a_{ij}| \, |t_j - l| + |l| \left|\sum_{j=1}^{\infty} a_{ij} - 1\right|.$$

The second term on the right hand side of $(*)$ is less than ϵM. With N fixed, by (ii) the first term is less than $MN\epsilon$ for large i and the last term is less than $|l| \epsilon$ for large i by (iii). Thus, $t \in c$ and $\lim At = l$.

We next give an application of the Uniform Boundedness Principle to a result in Fourier series. Let $f \in C_{\mathbb{C}}[-\pi, \pi]$. The nth partial sum of the Fourier series for f is given by

$$S_n(f)(t) = \frac{1}{2\pi} \sum_{j=-n}^{n} c_j(f)e^{ijt},$$

where $c_j(f) = \frac{1}{\sqrt{2\pi}} \int_{-\pi}^{\pi} f(s)e^{-ijs}ds$ is the jth Fourier coefficient of f with respect to the orthonormal set $\{\frac{1}{\sqrt{2\pi}}e^{ijt} : j \in \mathbb{Z}\}$ in $C_{\mathbb{C}}[-\pi, \pi]$. Thus, we have

$$S_n(f)(t) = \frac{1}{2\pi} \int_{-\pi}^{\pi} f(s) \sum_{j=-n}^{n} e^{ij(t-s)}ds.$$

The function $D_n(t) = \sum_{j=-n}^{n} e^{ijt}$ which appears in the integral above is called the *Dirichlet kernel*; we now compute a more useful form for the kernel. We have $(e^{it} - 1)D_n(t) = e^{i(n+1)t} - e^{-int}$ so $e^{-it/2}(e^{it} - 1)D_n(t) = e^{i(n+1/2)t} - e^{-i(n+1/2)t}$ which implies $D_n(t) = \sin(n+1/2)t / \sin(t/2)$. Thus,

$$S_n(f)(t) = \frac{1}{2\pi} \int_{-\pi}^{\pi} f(s)D_n(t - s)ds.$$

One of the principle questions for Fourier series is whether the Fourier series for the function converges to the function in some sense. We use the Uniform Boundedness Principle to show that that exists a continuous function whose Fourier series diverges at a point. Define a continuous linear functional F_n on $C_{\mathbb{C}}[-\pi, \pi]$ by

$$F_n(f) = S_n(f)(0) = \frac{1}{2\pi} \int_{-\pi}^{\pi} f(s)D_n(s)ds,$$

, i.e., $F_n(f)$ is the nth partial sum of the Fourier series for f evaluated at 0. Since $|F_n(f)| \leq \|f\|_\infty \frac{1}{2\pi} \int_{-\pi}^{\pi} |D_n(s)| ds$, F_n is continuous and $\|F_n\| \leq \frac{1}{2\pi} \int_{-\pi}^{\pi} |D_n(s)| ds$. We next show $\|F_n\| = \frac{1}{2\pi} \int_{-\pi}^{\pi} |D_n(s)| ds$. This equality follows from the following proposition.

Proposition 8.10. *Let $g : [a, b] \rightarrow \mathbb{R}$ be continuous. Define $G : C[a, b] \rightarrow \mathbb{R}$ by $G(f) = \int_a^b f(t)g(t)dt$. Then G is continuous and linear with $\|G\| = \int_a^b |g(t)| dt$.*

Proof: Since $|G(f)| \leq \|f\|_\infty \int_a^b |g(t)| dt$, G is continuous and $\|G\| \leq \int_a^b |g(t)| dt$. Fix $n \in \mathbb{N}$. Then

$$\int_a^b |g(t)| dt = \int_a^b |g| \frac{1 + n|g|}{1 + n|g|} = \int_a^b \frac{|g|}{1 + n|g|} + \int_a^b g \frac{ng}{1 + n|g|}$$

$$\leq \int_a^b 1/n + G(\frac{ng}{1 + n|g|}) \leq \frac{b - a}{n} + \|G\| \left\| \frac{ng}{1 + n|g|} \right\|$$

$$\leq \frac{b - a}{n} + \|G\|$$

which implies $\int_a^b |g(t)| dt \leq \|G\|$.

Next, we compute a lower bound for $\int_{-\pi}^{\pi} |D_n(s)| ds$. Since $\sin(u) \leq u$ for $0 \leq u \leq \pi$, we have

$$\int_{-\pi}^{\pi} |D_n(s)| ds \geq \int_0^{\pi} \left| \frac{\sin(2n+1)u}{u} \right| du = \sum_{j=0}^{2n} \int_{\frac{j\pi}{2n+1}}^{\frac{(j+1)\pi}{2n+1}} \left| \frac{\sin(2n+1)u}{u} \right| du$$

$$\geq \sum_{j=0}^{2n} \frac{2n+1}{(j+1)\pi} \int_{\frac{j\pi}{2n+1}}^{\frac{(j+1)\pi}{2n+1}} |\sin(2n + 1)u| du = \sum_{j=0}^{2n} \frac{1}{(j+1)\pi} \int_{j\pi}^{(j+1)\pi} |\sin u| du$$

$$= \frac{2}{\pi} \sum_{j=0}^{2n} \frac{1}{j+1}.$$

Thus, $\{\|F_n\|\}$ is unbounded. It follows from Theorem 2 that there exists a function $f \in C_{\mathbb{C}}[-\pi, \pi]$ such that $\{F_n(f)\}$ is unbounded, i.e., the Fourier series of f at 0 must diverge. Of course, the point 0 was only chosen for convenience.

Banach and Steinhaus developed an abstract result called the condensation of singularities which can be used to show the existence of a continuous function whose Fourier series diverges at any arbitrary countable subset of $[-\pi, \pi]$ (see [Sw1]6.6.31).

Likewise, there is an interesting application of the Uniform Boundedness Principle to interpolation. For this, let $0 = t_0 < t_1 < ... < t_n = 1$ be a collection of $n + 1$ nodes in $[0, 1]$ and define

$$l_i(t) = \frac{(t - t_0)...(t - t_{i-1})(t - t_{i+1})...(t - t_n)}{(t_i - t_0)...(t_i - t_{i-1})(t_i - t_{i+1})...(t_i - t_n)}$$

so $l_i(t_i) = 1$ and $l_i(t_j) = 0$ for $i \neq j$. For $f \in C[0, 1]$ the function

$$L(f)(t) = \sum_{i=0}^{n} f(t_i)l_i(t)$$

is called the *Lagrange interpolation polynomial* for f with respect to the nodes $\{t_i\}$, i.e., $L(f)$ is an nth order polynomial such that $L(f)(t_i) = f(t_i)$ for all i so $L(f)$ interpolates f at the t_i. Thus, if p is a polynomial of degree less than or equal to n, then, $L(p) = p$. If the function f has $n + 1$ continuous derivatives, the error in approximating f by $L(f)$ is of the order $\sup\{|f^{n+1}(t)| : 0 \le t \le 1\}/(n+1)!$ ([Na]). However, we use the Uniform Boundedness Principle to show that there exist continuous functions which are not well approximated by a sequence of interpolating polynomials. The map $f \to L(f)$ defines a linear operator L on $C[0, 1]$ and we show that L is continuous with $\|L\| = \sup\{\sum_{i=0}^{n} |l_i(t)| : 0 \le t \le 1\} = M$. First, $|L(f)(t)| \le \|f\|_{\infty} M$ implies that L is continuous with $\|L\| \le M$. For the other inequality, there exists s such that $M = \sum_{i=0}^{n} |l_i(s)|$ and now pick $f \in C[0, 1]$, $\|f\|_{\infty} = 1$, such that $f(t_i) = signl_i(s)$. Then $L(f)(s) = \sum_{i=0}^{n} f(t_i)l_i(s) = \sum_{i=0}^{n} |l_i(s)| = M$ so $\|L\| \ge M$. Now, it can be shown that $M \ge (\log n)/8\pi$ ([Na] p405); this is not easy). Thus, if $\tau^n = \{0 = t_0^n < t_1^n < ... < t_n^n = 1\}$ is any sequence of nodes, and L_n is the Lagrange interpolation operator with respect to τ^n, it follows from Theorem 2 that there exists $f \in C[0, 1]$ such that $\{L_n(f)\}$ does not converge in $C[0, 1]$.

We now give a sufficient condition for a sequence of positive operators on $C[0, 1]$ to converge to the identity operator. Let $\{S_n\}$ be a sequence of positive, linear operators on $C[0, 1]$; a linear operator $S : C[0, 1] \to C[0, 1]$

is positive if $S(f) \geq 0$ when $f \in C[0,1]$ and $f \geq 0$. Let $S_n : C[0,1] \to C[0,1]$ be a sequence of positive operators. We first establish a lemma. For $0 \leq u \leq 1$ let f_u be the function $f_u(t) = (t-u)^2$ for $0 \leq t \leq 1$.

Lemma 8.11. *If* $\lim S_n(1) = 1$, $\lim S_n(t) = t$ *and* $\lim S_n(t^2) = t^2$ *[convergence in* $(C[0,1], \|\cdot\|_\infty)$*], then* $\lim S_n(f_u)(u) = 0$ *uniformly for* $0 \leq u \leq 1$.

Proof: Let $\epsilon > 0$ and $0 \leq u \leq 1$. By expanding f_u it follws that $\|S_n(f_u) - f_u\|_\infty < \epsilon$ for large n. In particular, $|S_n(f_u)(u) - f_u(u)| = |S_n(f_u)(u)| < \epsilon$ for large n and $0 \leq u \leq 1$.

Theorem 8.12. *Suppose* $\lim S_n(1) = 1$, $\lim S_n(t) = t$ *and* $\lim S_n(t^2) = t^2$ *[convergence in* $(C[0,1], \|\cdot\|_\infty)$*]. Then* $\lim S_n(f) = f$ *for every* $f \in C[0,1]$ *[convergence in* $(C[0,1], \|\cdot\|_\infty)$*].*

Proof: Let $\epsilon > 0, f \in C[0,1]$. There exists $\delta > 0$ such that $|f(t) - f(u)| < \epsilon$ for $|t - u| < \delta$. Set $M = \|f\|_\infty$. We claim that if $0 \leq t \leq 1$, then

$$(\#) \quad -\epsilon - \frac{2M}{\delta^2}(t-u)^2 \leq f(t) - f(u) \leq \epsilon + \frac{2M}{\delta^2}(t-u)^2.$$

For, if $|t - u| < \delta$, then $-\epsilon < f(t) - f(u) < \epsilon$ while if $|t - u| \geq \delta$, then $-\frac{2M}{\delta^2}(t-u)^2 \leq -2M \leq f(t) - f(u) \leq 2M \leq \frac{2M}{\delta^2}(t-u)^2$.

For fixed u, from $(\#)$ and the positivity of the S_n we have

$$(*) \quad -\epsilon S_n(1) - \frac{2M}{\delta^2} S_n(f_u) \leq S_n(f) - f(u)S_n(1) \leq \epsilon S_n(1) + \frac{2M}{\delta^2} S_n(f_u).$$

Evaluating $(*)$ at u and using Lemma 11, we have $-2\epsilon \leq S_n(f)(u) - f(u) \leq 2\epsilon$ for large n or $\|S_n(f) - f\| \leq 2\epsilon$ for large n.

For a historical description of the history of the Uniform Boundedness Principle, see [Di], [Sw2].

Exercises.

1. Show that if the series $\sum_{j=1}^\infty s_j t_j$ converges for every $\{t_j\} \in c_0$, then $\{s_j\} \in l^1$ and defines a continuous linear functional with norm $\sum_{j=1}^\infty |s_j|$. Can c_0 be replaced by c_{00}?

2. Let X, Y, Z be normed spaces and $B : X \times Y \to Z$ be a bilinear mapping which is separately continuous, i.e., continuous in each variable separately. Show that if X is a Banach space, then B is jointly continuous, i.e., continuous on $X \times Y$ [recall Exercise 2.16 and use Theorem 2]. Show

the completeness is important (Hint: consider $B : c_{00} \times c_{00} \to \mathbb{R}$ defined by $B(s,t) = \sum_{j=1}^{\infty} s_j t_j$).

3. Let $F \subset L(X,Y)$, where X, Y are normed spaces. Show $\{\|T\| : T \in F\}$ is bounded iff F is uniformly bounded on bounded subsets of X iff F is equicontinuous at 0 iff F is equicontinuous on X. [F is equicontinuous at x if for every $\epsilon > 0$ there exists $\delta > 0$ such that $\|Tx - Ty\| < \epsilon$ for every $T \in F$ when $\|x - y\| < \delta$.]

4. Let $\{T_j\} \subset L(X,Y)$ be equicontinuous and Z a dense linear subspace of X. Show that if Y is complete and $\lim_j T_j z$ exists for every $z \in Z$, then $\lim_j T_j x = Tx$ exists for every $x \in X$ and $T \in L(X,Y)$.

5. Let $T_j \in L(X,Y)$ be such that $\{\|T_j\|\}$ is bounded. If $x_j \to 0$ in X, show $T_j x_j \to 0$. If $\lim_j T_j x = Tx$ for every $x \in X$, show $T \in L(X,Y)$ and $\lim_j T_j x = Tx$ uniformly for x belonging to compact subsets of X [Hint: use the first part and the sequential compactness].

6. Let X be the space of all real valued, continuous functions on $[0, \infty)$ which vanish at ∞ with the sup-norm. Show X is complete. If $a : [0, \infty) \to \mathbb{R}$ is such that the improper integral $\int_0^{\infty} a(t)x(t)dt$ converges for every $x \in X$, show $\int_0^{\infty} a(t)dt$ converges.

7. Show that a matrix $A = [a_{ij}]$ is conservative iff (i) $\sup_i \sum_{j=1}^{\infty} |a_{ij}| < \infty$ and (ii) for each p, $\lim i \sum_{j=p}^{\infty} a_{ij} = a_p$ exists.

8. Use the Baire Category Theorem to prove Theorem 2. (Hint: For each n let $X_n = \{x \in X : \|Tx\| \le n \ \forall \ T \in \Gamma\}$.)

9. Show linear interpolation satisfies the conditions of Theorem 11. Show Lagrange interpolation does not satisfy the conditions.

Chapter 9

Weak Convergence

In this chapter we define weak and weak* convergence in normed spaces and give several examples and applications. Let X be a normed space.

Definition 9.1. A sequence $\{x_j\}$ in a normed space is weakly convergent to $x \in X$ if $f(x_j) \to f(x)$ for every $f \in X'$.

Thus, if $\{x_j\}$ converges to x with respect to the norm in X, then $\{x_j\}$ converges weakly to x. The converse is false. See, however, the Hahn-Schur Theorem later in the chapter.

Example 9.2. In c_0, the sequence $\{e^j\}$ converges weakly to 0, but does not converge in norm.

Proposition 9.3. If $\{x_j\}$ converges weakly to $x \in X$ and $y \in X$, then $x = y$; that is, weak limits are unique.

Proof: Suppose $x \neq y$. Then there exists $f \in X'$ such that $f(x) \neq f(y)$ [Corollary 7.4]. But, $\lim_j f(x_j) = f(x) = f(y)$ giving a contradiction.

Proposition 9.4. If $\{x_j\}$ converges weakly to x, then $\{\|x_j\|\}$ is bounded and $\|x\| \leq \liminf \|x_j\|$.

Proof: Let J be the canonical imbedding of X into X''. Then $J(x_j)(f) \to J(x)(f)$ for every $f \in X'$. The result follows from Corollary 8.4 and Theorem 8.6.

Proposition 9.5. Let $T \in L(X,Y)$. If $\{x_j\}$ converges weakly to x, then $\{Tx_j\}$ converges weakly to Tx.

Proof: Let $g \in Y'$. Then $gT \in X'$ so $gT(x_j) \to gT(x)$.

Theorem 9.6. *Let X be reflexive. If $\{x_j\}$ is bounded in X, then $\{x_j\}$ has a weakly convergent subsequence.*

Proof: Let X_1 be the closed span of $\{x_j\}$. Then X_1 is separable and reflexive by Theorem 7.20 so X_1' is separable by Corollary 7.16. Let $\{x_j'\}$ be dense in X_1'. Now $\{x_i : i \in \mathbb{N}\}$ is pointwise bounded on $\{x_j' : j \in \mathbb{N}\}$ so by the diagonalization procedure, there is a subsequence $\{n_i\}$ such that $\{x_{n_i}\}$ converges pointwise on $\{x_j' : j \in \mathbb{N}\}$, i.e., $\{x_j'(x_{n_i})\}_i$ converges for every j ([DeS] 25.10). We claim that $\{x'(x_{n_i})\}_i$ converges for every $x' \in X_1'$. Let $\epsilon > 0$. Since $\{x_j'\}$ is dense in X_1', there exists j such that $\|x_j' - x'\| < \epsilon/3M$, where $M > 0$ is such that $\|x_i\| \leq M$ for every i. There exists N such that $k, l \geq N$ implies $\left|x_j'(x_{n_k} - x_{n_l})\right| < \epsilon/3$. Then

$$|x'(x_{n_k}) - x'(x_{n_l})| \leq \|x' - x_j'\|\|x_{n_k}\| + \|x' - x_j'\|\|x_{n_l}\| + |x_j'(x_{n_k} - x_{n_l})| < \epsilon.$$

Hence, $\{x'(x_{n_i})\}_i$ converges. Define $x'' : X_1' \to \Bbbk$ by $x''(x') = \lim_i x'(x_{n_i})$. Then $x'' \in X_1''$ by the Banach-Steinhaus Theorem 8.4. Since X_1' is reflexive, there exists $x \in X_1$ such that $J_{X_1} x(x') = x'(x) = x''(x') = \lim_i x'(x_{n_i})$ for every $x' \in X_1'$.

We show $\{x_{n_i}\}$ converges weakly to x in X. Let $z' \in X'$ and let z_1' be the restriction of z to X_1. So $z_1' \in X_1'$ and

$$z'(x) = z_1'(x) = x''(z_1') = \lim_i z_1'(x_{n_i}) = \lim_i z'(x_{n_i}).$$

Example 9.7. Reflexivity in Theorem 6 is important. Consider $\{e^j\}$ in l^1.

Definition 9.8. A sequence $\{x_j\}$ in a normed space X is weakly Cauchy if the sequence $\{x'(x_j)\}$ converges for every $x' \in X'$. A normed space X is weakly sequentially complete if every weak Cauchy sequence is weakly convergent.

Example 9.9. c_0 is not weakly sequentially complete. Let $x^n = \sum_{j=1}^n e^j$. Then $\lim x'(x^n)$ exists for every $x' \in c_0' = l^1$ but there exists no $x \in c_0$ such that $\{x^n\}$ converges weakly to x.

Proposition 9.10. *A reflexive space X is weakly sequentially complete.*

Proof: Suppose $\{x_j\} \subset X$ is such that $\lim x'(x_j)$ exists for every $x' \in X'$. Then $\{\|x_j\|\}$ is bounded by Theorem 8.6. By Theorem 6 there exists a subsequence $\{x_{n_j}\}$ which converges weakly to some $x \in X$. Then $\lim_j x'(x_{n_j}) = x'(x) = \lim x'(x_i)$ for every $x' \in X'$.

Thus, Example 9 shows that c_0 is not reflexive. We will show later in Corollary 15 that the converse of Proposition 10 is false $(X = l^1)$.

We now establish an interesting and useful result due to Hahn and Schur which in particular shows that weak and norm convergence for sequences coincide in l^1. We first need two lemmas.

Lemma 9.11. *If $t_j \in \mathbb{R}$ and there exists $M > 0$ such that $\left|\sum_{j \in \sigma} t_j\right| \leq M$ for every finite $\sigma \subset \mathbb{N}$, then $\{t_j\} \in l^1$ and $\sum_{j=1}^{\infty} |t_j| \leq 2M$.*

Proof: For $\sigma \subset \mathbb{N}$, let $\sigma_+ = \{j \in \sigma : t_j \geq 0\}$ and $\sigma_- = \{j \in \sigma : t_j < 0\}$. Then $\sum_{j \in \sigma_+} |t_j| = \sum_{j \in \sigma_+} t_j \leq M$ and $\sum_{j \in \sigma_-} |t_j| = -\sum_{j \in \sigma_-} t_j \leq M$. So $\sum_{j \in \sigma} |t_j| \leq 2M$. Since σ is arbitrary, $\sum_{j=1}^{\infty} |t_j| \leq 2M$.

Lemma 9.12. *Let $a^i = \{a_{ij}\}_j \in l^1$ for every $i \in \mathbb{N}$, suppose $\lim_i a_{ij} = a_j$ exists for every j and set $F = \{\sigma \subset \mathbb{N} : \sigma \text{ finite}\}$. If $\lim_i \sum_{j \in \sigma} a_{ij} = \sum_{j \in \sigma} a_j$ is not uniform for $\sigma \in F$, then there exists $\epsilon > 0$, an increasing sequence $\{i_j\}$, an increasing sequence $\{\tau_j\} \subset F$ such that $\left|\sum_{k \in \tau_j} (a_{i_j k} - a_k)\right| > \epsilon$.*

Proof: If not, there exists $\epsilon > 0$ such that for every i there exist k_i and finite σ_i such that $\left|\sum_{k \in \sigma_i} (a_{k_i k} - a_k)\right| > 2\epsilon$. Put $i_1 = 1$ and let $k_1, \sigma_1 \in F$ be such that $\left|\sum_{k \in \sigma_1} (a_{k_1 k} - a_k)\right| > 2\epsilon$. Set $\tau_1 = \sigma_1$ and $M_1 = \max \sigma_1$. There exists n_1 such that $\sum_{j=1}^{M_1} |a_{ij} - a_j| < \epsilon$ for $i \geq n_1$. There exists $i_2 > \max\{i_1, n_1\}$ and $\sigma_2 \in F$ such that $\left|\sum_{k \in \sigma_2} (a_{i_2 k} - a_k)\right| > 2\epsilon$. Set $\tau_2 = \sigma_2 \backslash \sigma_1$ and note

$$\left|\sum_{k \in \tau_2} (a_{i_2 k} - a_k)\right| \geq \left|\sum_{k \in \sigma_2} (a_{i_2 k} - a_k)\right| - \sum_{k \in \sigma_1} |a_{i_2 k} - a_k| > 2\epsilon - \epsilon.$$

Now just continue.

Theorem 9.13. *(Hahn-Schur) Let $a^i = \{a_{ij}\}_j \in l^1$ for every $i \in \mathbb{N}$ and suppose $\lim_i \sum_{j \in \sigma} a_{ij} = 0$ for every $\sigma \subset \mathbb{N}$. Then $\lim_i \sum_{j=1}^{\infty} |a_{ij}| = \lim_i \|a^i\|_1 = 0$.*

Proof: First, we claim that $\lim_i \sum_{j \in \sigma} a_{ij} = 0$ uniformly for finite σ. If this fails, let the notation be as in the preceding lemma. Define an infinite matrix

$$Z = [z_{pq}] = \left[\sum_{k \in \tau_q} a_{i_p k}\right].$$

The columns of Z go to 0 by hypothesis and the rows of Z go to 0 by the convergence of the series so we may apply Lemma 8.1. Thus, there exists an increasing sequence $\{m_p\}$ such that $\left|z_{m_p m_q}\right| < \epsilon/2^{p+q}$ for $p \neq q$. Set $\tau = \cup_{q=1}^{\infty} \tau_{m_q}$ Then

$$\left|\sum_{q \in \tau} a_{i_p q}\right| = \left|\sum_{q=1}^{\infty} \sum_{k \in \tau_{m_q}} a_{i_p k}\right|$$

$$\geq \left|\sum_{k \in \tau_{m_p}} a_{i_p k}\right| - \sum_{q \neq p} \left|\sum_{k \in \tau_{m_q}} a_{i_p k}\right| \geq \epsilon - \sum_{q \neq p} \epsilon/2^{p+q} > \epsilon/2$$

which contradicts the hypothesis.

In particular, the Hahn-Schur Theorem above implies that weakly convergent sequences in l^1 are norm convergent. For those readers familiar with topology this example is sometimes used to show that sequences are not adequate to define topologies; there is a topology on any normed space, called the weak topology, such that convergence for sequences in the weak topology is exactly weak sequential convergence. If the normed space is infinite dimensional, the weak topology is always strictly weaker than the norm topology so sequential convergence cannot be used to define the weak topology of l^1.

We have a stronger result than that given in the Hahn-Schur Theorem.

Corollary 9.14. *Let $a^i = \{a_{ij}\}_j \in l^1$ for every $i \in \mathbb{N}$ and assume that $\lim_i \sum_{j \in \sigma} a_{ij}$ exists for every $\sigma \subset \mathbb{N}$ with $a_j = \lim_i a_{ij}$. Then $a = \{a_j\} \in l^1$ and*

$$\lim_i \sum_{j=1}^{\infty} |a_{ij} - a_j| = \lim_i \left\|a^i - a\right\|_1 = 0.$$

Proof: Let $\{m_i\}, \{n_i\}$ be increasing sequences with $m_i < n_i < m_{i+1}$. Then $\lim_i \sum_{j \in \sigma} (a_{m_i j} - a_{n_i j}) = 0$ for every $\sigma \subset \mathbb{N}$. By the Hahn-Schur Theorem, $\lim_i \sum_{j=1}^{\infty} |a_{m_i j} - a_{n_i j}| = 0$. If $\epsilon > 0$, there exists i_0 such that $i \geq i_0$ implies $\sum_{j=1}^{\infty} |a_{m_i j} - a_{n_i j}| < \epsilon$ so $\left|\sum_{j \in \sigma} (a_{m_i j} - a_{n_i j})\right| < \epsilon$ for every $\sigma \subset \mathbb{N}$. For finite σ, $\left|\sum_{j \in \sigma} (a_{m_i j} - a_j)\right| \leq \epsilon$ if $i \geq i_0$. By Lemma 11, $\sum_{j=1}^{\infty} |a_{m_i j} - a_j| \leq 2\epsilon$ so $a = \{a_j\} \in l^1$ and $\lim_i \left\|a^{m_i} - a\right\|_1 = 0$. Since $\{m_i\}$ is arbitrary, $\lim_i \left\|a^i - a\right\|_1 = 0$.

Corollary 9.15. *l^1 is weakly sequentially complete; moreover, if $\{a^i\}$ is a weak Cauchy sequence in l^1, then there exists $a \in l^1$ such that $\lim_i \left\|a^i - a\right\|_1 = 0$.*

We can use Corollary 14 to establish a summability result concerning regular matrices due to Steinhaus.

Theorem 9.16. *(Steinhaus) Let $A = [a_{ij}]$ be a regular matrix. Then there exists $t \in m_0$ such that $At \notin c$; that is there is a sequence in m_0 which is not summable by A.*

Proof: Suppose $A : m_0 \rightarrow c$. Then $\lim_i \sum_{j=1}^{\infty} a_{ij} t_j$ exists for every $t \in m_0$. Let $a_j = \lim_i a_{ij}$. By Corollary 14, $\lim_i \sum_{j=1}^{\infty} a_{ij} t_j = \sum_{j=1}^{\infty} a_j t_j$ for every $t \in m_0$. Set $t^n = \sum_{j=n}^{\infty} e^j$. So $\lim_i \sum_{j=n}^{\infty} a_{ij} = \sum_{j=n}^{\infty} a_j \rightarrow 0$ as $n \rightarrow \infty$ by the convergence of the series. But, since A is regular, $\lim_i \sum_{j=n}^{\infty} a_{ij} = 1$ for every n giving a contradiction.

Another type of weak convergence in dual spaces is called weak* convergence.

Definition 9.17. A sequence $\{f_j\}$ in the dual X' of a normed space X is weak* convergent to $f \in X'$ if $f_j(x) \rightarrow f(x)$ for every $x \in X$.

Of course, if X is reflexive, then a sequence $\{f_j\}$ is weak* to f iff $\{f_j\}$ is weakly convergent to f. The sequence $\{e^j\}$ in $l^1 = (c_0)'$ is weak* convergent to 0 but is not weakly convergent. We have the following property of weak* convergent sequences in the duals of Banach spaces.

Theorem 9.18. *Let X be a Banach space and $\{f_j\} \subset X'$. If $\{f_j\}$ is weak* convergent to $f \in X'$, then $\{\|f_j\|\}$ is bounded.*

Proof: This follows directly from the Uniform Boundedness Principle 8.2.

Completeness in the result above is important.

Example 9.19. The sequence $\{je^j\}$ in $l^1 = (c_{00})'$ is weak* convergent to 0 but $\|je^j\| = j$.

As an example of weak* convergence we consider the approximation of the definite integral of a continuous function. If $f \in C[a,b]$, we write $I(f) = \int_a^b f$ and note that I is a continuous linear functional on $C[a,b]$ with $\|I\| = b - a$ (Exercise 5). To approximate $I(f)$ we consider a finite set of nodes $a = t_0 < t_1 < ... < t_n = b$ and coefficients $c_0, c_1, ..., c_n$ and use $S(f) = \sum_{j=0}^{n} c_j f(t_j)$ to approximate $I(f)$; of course, S depends on the nodes $\{t_j\}$ and the coefficients $\{c_j\}$. Simpson's rule and the trapezoidal rule are examples of such an approximation scheme. S is obviously a linear functional on $C[a, b]$ and we show that S is continuous with $\|S\| = \sum_{j=1}^{n} |c_j|$. First, we have $|S(f)| \leq \|f\|_\infty \sum_{j=1}^{n} |c_j|$ so $\|S\| \leq \sum_{j=1}^{n} |c_j|$. Next, pick a continuous function f on $[a, b]$ with $f(t_j) = signc_j$ $(j = 0, 1, ..., n)$ and $\|f\|_\infty = 1$ so $S(f) = \sum_{j=1}^{n} |c_j|$ and, therefore, $\|S\| = \sum_{j=1}^{n} |c_j|$.

To approximate $I(f)$ we consider a sequence of nodes $\tau^n = \{a = t_0^n < t_1^n < ... < t_n^n = b\}$ and coefficients $\{c_j^n : j = 0, 1, ..., n\}$ and set $S_n(f) = \sum_{j=1}^{n} c_j^n f(t_j^n)$. To ask that $\{S_n(f)\}$ approximate $I(f)$ for every $f \in C[a, b]$ means that we want the sequence $\{S_n\}$ to converge weak* to I in $C[a, b]'$. For this we have a criterion due to Szegö.

Theorem 9.20. $\{S_n\}$ *is weak* convergent to I iff (i) $M = \sup\{\|S_n\| : n \in \mathbb{N}\} < \infty$ and (ii) $S_n(p) \to I(p)$ for every polynomial p.*

Proof: Suppose $S_n \to I$ weak*. Then (i) follows from 8.2 and (ii) is clear.

Since the polynomials are dense in $C[a, b]$, the converse follows from Exercise 8.4.

Of course, any dense subset of $C[a, b]$ could be used in the theorem in place of the polynomials.

We consider the possibility of choosing functionals $\{S_n\}$ satisfying condition (ii) with $[a, b] = [0, 1]$. With a fixed set of nodes $0 = t_0 < t_1 < ... < t_n = 1$ we show that the coefficients $\{c_j\}$ can be chosen so that $S(p) = I(p)$ for any polynomial p of degree less than or equal to n. Let $L(f)(t) = \sum_{i=0}^{n} f(t_i)l_i(t)$ be the Lagrange interpolation functional with respect to the nodes $\{t_0 < ... < t_n\}$ so that $L(p) = p$ for every polynomial p with degree $\leq n$ (see the final paragraphs in Chapter 8). If we consider approximating f by $L(f)$ and, then, $I(f)$ by $\int_0^1 L(f)(t)dt$ we obtain

$$\int_0^1 L(f)(t)dt = \sum_{i=0}^{n} f(t_i) \int_0^1 l_i(t)dt = \sum_{i=0}^{n} c_i f(t_i) = S(f)$$

for the approximating scheme and we have that $I(p) = S(p)$ for every polynomial of degree $\leq n$. Now, choose a sequence of nodes $\tau^n = \{t_0^n < ... < t_n^n\}$ and let S_n be the approximating functional constructed above for τ^n. Then condition (ii) is trivially satisfied for the sequence $\{S_n\}$.

Concerning condition (i), we have

Theorem 9.21. *If the coefficients $\{c_j^n : j = 0, 1, ..., n\}$ are all non-negative and condition (ii) is satisfied, $\sup\{\|S_n\| : n \in \mathbb{N}\} \leq b - a$.*

Proof: $\lim_n S_n(1) = \lim_n \sum_{j=0}^{n} c_j^n = \lim_n \sum_{j=0}^{n} |c_j^n| = \lim_n \|S_n\| = I(1) = b - a$.

It is possible to satisfy conditions (i) and (ii). For this one chooses the nodes $\tau^n = \{t_0^n < \dots < t_n^n\}$ such that the polynomials $\{w_n(t) = (t - t_0^n)\dots(t - t_n^n)\}$ are orthogonal and then the hypothesis in the theorem above is satisfied. The resulting quadrature formulas are said to be of Gaussian type. This type of construction is described in [Na].

Exercises.

1. Show that in a finite dimensional normed space a sequence converges weakly iff it converges in norm.

2. Show that in a normed space a sequence $\{x_j\}$ is Cauchy iff for every pair of increasing sequences $\{n_j\}, \{m_j\}$ with $m_j < n_j < m_{j+1}$ we have $x_{m_j} - x_{n_j} \to 0$.

3. Let $1 < p < \infty$. Give an example of a sequence in l^p which converges weakly but not in norm.

4. Let X be a normed space with X' separable. Show that if $\{x_j\} \subset X$ is bounded, there exists a subsequence $\{x_{n_j}\}$ which is weakly Cauchy. [Hint: Use the diagonalization procedure ([DeS] 25.10).]

5. Show the linear functional $I(f) = \int_a^b f$ is continuous on $C[a,b]$ with $\|I\| = b - a$.

6. Let $1 \le p < \infty$. Show a sequence $t^k = \{t_j^k\}_{j=1}^{\infty}$ in l^p converges weakly to 0 iff $\{\|t^k\|_p\}$ is bounded and $\lim_k t_j^k = 0$ for every j. Give a similar characterization for weak convergence in c_0.

7. Show $\{e^j\}$ is not weak* convergent in $(l^{\infty})'$ and has no weak* convergent subsequences.

8. In a Hilbert space H, show $\{x_j\}$ converges to x iff $\{x_j\}$ converges weakly to x and $\|x_j\| \to \|x\|$.

Chapter 10

The Open Mapping and Closed Graph Theorems

In this chapter we prove the third of the three fundamental principles of functional analysis, the open mapping/closed graph theorems. We give a proof using an interesting lemma due to Zabreiko.

Definition 10.1. A semi-norm p on a normed space X is countably subadditive if $p(\sum_{j=1}^{\infty} x_j) \le \sum_{j=1}^{\infty} p(x_j)$ for every convergent series $\sum_{j=1}^{\infty} x_j$.

Example 10.2. Define a semi-norm p on c_{00} by $p(t) = p(\{t_j\}) = \sum_{j=1}^{\infty} |t_j|$ (finite sum). We claim p is countably subadditive. Suppose $t = \sum_{k=1}^{\infty} t^k$ converges in c_{00}. Then $t_j = \sum_{k=1}^{\infty} t_j^k$ for every j and $p(t) = \sum_{j=1}^{\infty} |t_j| = \sum_{j=1}^{\infty} \left| \sum_{k=1}^{\infty} t_j^k \right| \le \sum_{j=1}^{\infty} \sum_{k=1}^{\infty} |t_j^k| = \sum_{k=1}^{\infty} \sum_{j=1}^{\infty} |t_j^k| = \sum_{k=1}^{\infty} p(t^k)$.

We now state and prove Zabreiko's lemma. The proof uses the Baire Category Theorem which is presented in Appendix D.

Lemma 10.3. *(Zabreiko) If X is a Banach space, then every countably subadditive semi-norm p on X is continuous.*

Proof: It suffices to show that p is continuous at 0 (Proposition 1.3). Let $\epsilon > 0$. Put $E = \{x : p(x) \le \epsilon\}$. Note that E is symmetric and convex and $X = \cup_{n=1}^{\infty} nE$. By the Baire Category Theorem (Appendix D), some $\overline{nE}$ contains a ball so $\overline{E}$ contains a ball $S(x, r) = \{y : \|x - y\| < r\}$. Since E is symmetric, $\overline{E}$ contains $-S(x, r)$ and since E is convex, $\frac{1}{2}S(x,r) - \frac{1}{2}S(x,r) \subset \overline{E}$ so there exists an open ball $S(0, \delta) \subset \overline{E}$.

We claim that $z \in S(0, \delta)$ implies $p(z) \le 2\epsilon$ which will show p is continuous at 0. Let $z \in S(0, \delta) \subset \overline{E}$ so there exists $x_1 \in E$ such that $\|z - x_1\| < \delta/2$. Then $z - x_1 \in \frac{1}{2}S(0, \delta) \subset \frac{1}{2}\overline{E}$ so there exists $x_2 \in \frac{1}{2}E$ such that $\|z - x_1 - x_2\| < \frac{1}{2^2}\delta$. Continuing this construction, there is a sequence $\{x_j\}$ such that $x_j \in \frac{1}{2^{j-1}}E$ and $z - \sum_{j=1}^{k} x_j \in \frac{1}{2^k} S(0, \delta)$. Hence,

$p(x_j) \leq \frac{1}{2^{j-1}}\epsilon$ and $z = \sum_{j=1}^{\infty} x_j$. Since p is countably subadditive, $p(z) \leq \sum_{j=1}^{\infty} p(x_j) \leq \sum_{j=1}^{\infty} \frac{1}{2^{j-1}}\epsilon = 2\epsilon$.

We now use Zabreiko's Lemma to prove the Closed Graph Theorem. Let X, Y be normed spaces and let A be a linear operator with domain $\mathcal{D}(A) \subset X$ and range in Y. The *graph* of A is $G(A) = \{(x, Ax) : x \in \mathcal{D}(A)\}$.

Definition 10.4. A is closed if the graph of A, $G(A)$, is closed in $X \times Y$ (here we use the norm $\|(x, y)\| = \|x\| + \|y\|$ on $X \times Y$).

Proposition 10.5. *A is closed iff whenever $x_j \in \mathcal{D}(A)$, $x_j \to x$ and $Ax_j \to y$, then $(x, y) \in G(A)$ and $Ax = y$.*

Proof; Just note that a sequence $\{(x_j, y_j)\} \subset X \times Y$ converges to $(x, y) \in X \times Y$ iff $x_j \to x$ and $y_j \to y$.

Corollary 10.6. *If X, Y are normed spaces and $T \in L(X, Y)$, then T is closed.*

The following examples compare closed and continuous linear operators.

Example 10.7. Let $X = Y = C[0, 1]$. Define $D : C^1[0, 1] \to X$ by $Df = f'$. Then D is linear but not continuous. However, D is closed. For suppose $f_j \to f$ in X, $f_j \in C^1[0, 1]$ and $Df_j \to g$. Then $f \in C^1[0, 1]$ and $f' = Df = g$ ([DeS] 11.7).

Example 10.8. Let $X = Y = c_0$. The identity operator $I : c_{00} \to X = c_0$ is continuous but not closed.

This example shows that in Corollary 6, it is important that the domain be the entire space.

We now establish the Closed Graph Theorem.

Theorem 10.9. *(Closed Graph Theorem) Let X, Y be Banach spaces and $T : X \to Y$ be a linear operator with a closed graph. Then T is continuous.*

Proof: Define a semi-norm p on X by $p(x) = \|Tx\|$. We claim that p is countably subadditive. Suppose $\sum_{j=1}^{\infty} x_j = x$ is convergent in X and we may assume $\sum_{j=1}^{\infty} p(x_j) < \infty$. Then $\sum_{j=1}^{\infty} Tx_j$ is absolutely convergent in Y and , therefore, converges to some $y \in Y$. By Proposition 5, $y = Tx = T(\sum_{j=1}^{\infty} x_j)$ so $p(x) = \|Tx\| \leq \sum_{j=1}^{\infty} \|Tx_j\| = \sum_{j=1}^{\infty} p(x_j)$. Thus, p is countably subadditive and continuous by Zabreiko's Lemma which implies that T is continuous.

Example 7 shows that the completeness of the domain space in the Closed Graph Theorem is important even when the range space is complete. We give an example to show that the completeness of the range space is important even when the domain space is complete. For this, we need an observation.

Proposition 10.10. *If* $A : \mathcal{D}(A) \subset X \to Y$ *is linear, closed and 1-1, then* A^{-1} *is closed.*

Proof: $G(A^{-1}) = \{(Ax, x) : x \in \mathcal{D}(A)\}$.

Example 10.11. Let X be a separable Banach space with a Hamel basis $\{x_h : h \in H\}$ with $\|x_h\| = 1$ for $h \in H$ (recall H is uncountable, Exercise 4.6). Define a norm $\|\cdot\|'$ on X by $\|x\|' = \sum_{h \in H} |t_h|$, where $x = \sum_{h \in H} t_h x_h$. Since $\|x_h\| = 1$, $\|x\| \le \sum_{h \in H} |t_h| = \|x\|'$ and the identity $I : (X, \|\cdot\|') \to (X, \|\cdot\|)$ is continuous (and closed, Corollary 6). However, since $(X, \|\cdot\|')$ is not separable ($\|x_h - x_k\|' = 2$ for $h \ne k$), $I : (X, \|\cdot\|) \to (X, \|\cdot\|')$ is not continuous but has a closed graph by the Proposition above.

The utility of the Closed Graph Theorem is that it often gives a useful way to check for continuity of a linear operator. We now use the Closed Graph Theorem to establish the Open Mapping Theorem. We first consider a special case.

Theorem 10.12. *Let* X, Y *be Banach spaces. If* $T : X \to Y$ *is linear, continuous and 1-1, onto, then the inverse operator* $T^{-1} : Y \to X$ *is continuous, i.e.,* T *is a homeomorphism.*

Proof: T^{-1} is closed by Proposition 10 so the result follows from the Closed Graph Theorem.

Theorem 10.13. *(Open Mapping Theorem) Let* X, Y *be Banach spaces and* $T : X \to Y$ *be linear, continuous and onto. Then* T *is an open mapping, i.e.,* T *carries open sets to open sets.*

Proof: Let $\hat{T} : X/\ker T \to Y$ be the induced operator (Chapter 3). Then $X/\ker T$ is a Banach space, $\hat{T}$ is linear, continuous, 1-1 (Chapter 3). By Theorem 12, $\hat{T}$ is a homeomorphism. Since the quotient map $X \to X/\ker T$ is open (Chapter 3), the result follows.

Again as in the Closed Graph Theorem, the completeness of both domain and range spaces is important (see Exercise 2).

We derive several consequences for the Open Mapping Theorem. First, from Theorem 12, we have

Proposition 10.14. *Let X be a vector space which is complete with respect to two norms, $\|\cdot\|$ and $\|\cdot\|'$. Suppose the identity $I : (X, \|\cdot\|) \to (X, \|\cdot\|')$ is continuous. Then the two norms are equivalent.*

Again the completeness in Proposition 14 is important (Exercise 3).

In some expositions the Closed Graph Theorem is derived from the Open Mapping Theorem; see Maddox ([Ma]). The Open Mapping Theorem can also be derived directly from Zabreiko's Lemma (see Megginson ([Me]).

In the next two chapters we give applications of the Closed Graph/Open Mapping Theorems to projections and Schauder bases.

Exercises.

1. Give an example showing that the completeness is important in Zabreiko's Lemma.

2. Give examples showing that the completeness assumptions in the Open Mapping Theorem are important even in the presence of completeness of the other space.

3. Give an example of a normed space X with two norms $\|\cdot\|$ and $\|\cdot\|'$ such that $(X, \|\cdot\|)$ is complete, the identity map $I : (X, \|\cdot\|) \to (X, \|\cdot\|')$ is continuous but the inverse map is not. (Hint: Consider $C[0,1]$.)

4. Let X, Y be Banach spaces and $T : X \to Y$ be linear. If there exists $\{y'_a : a \in A\} \subset Y'$ which separates the points of Y and $y'_a T$ is continuous for every $a \in A$, show T is continuous.

5. Let X, Y be Banach spaces and $T : X \to Y$ be linear. The graph norm on X is $\|x\|' = \|x\| + \|Tx\|$. Show $\|\cdot\|'$ is complete iff T is closed.

6. Suppose $\|\cdot\|$ is a complete norm on $C[0,1]$ such that $\|f_j - f\| \to 0$ implies $f_j(t) \to f(t)$ for every $t \in [0,1]$. Show $\|\cdot\|$ is equivalent to $\|\cdot\|_\infty$.

7. Let $1 \le p, q < \infty$. If $A = [a_{ij}]$ is an infinite matrix which maps l^p into l^q, show A is continuous. (Hint: Use the Closed Graph Theorem.) What about similar matrices with domain or range c_0?

8. Use Zabreiko's Lemma to derive the Uniform Boundedness Theorem 8.2. (Hint: Consider the semi-norm $p(x) = \sup_{T \in \Gamma} \|Tx\|$.)

Chapter 11

Projections

If X is a vector space and M is a linear subspace, there is always a subspace N such that $X = M \oplus N$. In analysis, it is important to know when X is a Banach space and M is a closed subspace, is there a closed subspace N such that $X = M \oplus N$? This is the case when X is a Hilbert space, we just need to take $N = M^\perp$ (5.17). However, as is shown in Theorem 8 this is not the case when X is a general Banach space. In this chapter we will consider this problem.

Let X be a Banach space.

Definition 11.1. A projection in X is a continuous linear operator $P : X \to X$ such that $P^2 = P$.

Note that if P is a projection, then $I - P$ is also a projection, where I is the identity operator.

Proposition 11.2. *Let P be a projection in X. Set $M = \{x : Px = x\}$ and $N = \{x : Px = 0\} = \{x : (I - P)x = x\}$. Then M and N are closed linear subspaces with $X = M \oplus N$.*

Proof: Clearly M and N are closed subspaces and $M \cap N = \{0\}$. Let $x \in X$. Then $x = Px + (I - P)x$. Now $Px \in M$ since $P(Px) = Px$ and $(I - P)x \in N$ since $P(I - P)x = (P - P^2)x = Px - Px = 0$.

Note in Proposition 2 that $\mathcal{R}P = M$, where $\mathcal{R}P$ denotes the range of P.

Definition 11.3. Two closed subspaces M and N of X are complementary if $X = M \oplus N$.

Proposition 11.4. *Let M and N be closed, complementary subspaces of X. For $x = m + n$ with $m \in M$, $n \in N$ define $Px = m$. Then P is a projection with $\mathcal{R}P = M$ and $\ker P = N$.*

Proof: Clearly $P^2 = P$, P is linear, $\mathcal{R}P = M$ and $\ker P = N$. We only need to show that P is continuous. By the Closed Graph Theorem, it suffices to show that P is closed. Suppose $x_j \to x$ and $Px_j \to y$ in X. Then $Px_j \in M$ implies $y \in M$ since M is closed. Now $x_j = Px_j + (I - P)x_j$ and since $\{Px_j\}$ and $\{x_j\}$ both converge, $(I - P)x_j \to z \in X$. Therefore, $z \in N$ since N is closed and $x = y + z$, where $y \in M$ and $z \in N$. Hence, $Px = y$ and P is closed.

Remark 11.5. From Propositions 2 and 4 it follows that there is a one-to-one correspondence between projections in X and complemented subspaces.

As noted earlier if M is a closed subspace of a Banach space X, there may not exist a projection from X onto M (Theorem 8). However, we do have

Proposition 11.6. *Let M be a finite dimensional subspace of X. Then there exists a projection P from X onto M.*

Proof: Let $\{x_1, ..., x_n\}$ be a basis for M. We construct $\{f_1, ..., f_n\} \subset X'$ such that $f_i(x_j) = \delta_{ij}$. Fix i and set $M_i = span\{x_j : j \neq i\}$. Then M_i is closed and $x_i \notin M_i$. By Theorem 7.2, there exists $g_i \in X'$ such that $g_i(x_i) \neq 0$ and $g_i(M_i) = \{0\}$. Set $f_i = g_i/g_i(x_i)$.
Now define P by $Px = \sum_{i=1}^{n} f_i(x)x_i$.

We show that the space c_0 is not complemented in l^∞. The proof we give is due to Whitley ([Wh]).

Lemma 11.7. *Let I be a countable set. There exists a family $\{u_a : a \in A\}$ of subsets of I such that (1) u_a is infinite, (2) $u_a \cap u_b$ is finite for $a \neq b$ and (3) A is uncountable.*

Proof: (Art Kruse) Take I to be the rationals in $(0, 1)$, A the irrationals in $(0, 1)$ and , for $a \in A$, let u_a be a sequence of rationals converging to a.

We say a Banach space X has the *countable total subset property* (CTS) if the dual X' has a countable set which separates the points of X. It is clear that CTS is preserved under topological isomorphisms, subspaces of spaces with CTS have CTS and l^∞ has CTS (consider $\{e^j\}$ in the dual of l^∞).

Theorem 11.8. *There exists no continuous projection from l^∞ onto c_0.*

Proof: Suppose there exists a continuous projection from l^∞ onto c_0. Then there is a closed subspace M such that $l^\infty = c_0 \oplus M$ so l^∞/c_0 is topologically isomorphic to M and, therefore, l^∞/c_0 must have CTS by the observations above. We show that this is not possible. Let $\pi : l^\infty \to l^\infty/c_0$ be the quotient map and let $\{u_a : a \in A\}$ be as in Lemma 7 with $I = \mathbb{N}$.

For $g \in (l^\infty/c_0)'$ we claim that $C = \{\pi\chi_{u_a} : g(\pi\chi_{u_a}) \neq 0\}$ is countable. For this it suffices to show $C_n = \{\pi\chi_{u_a} : |g(\pi\chi_{u_a})| \geq 1/n\}$ is finite for every n. Choose $\pi\chi_{u_{a_1}} = f_1, ..., \pi\chi_{u_{a_m}} = f_m$ from C_n and set $b_i = signg(f_i)$ and $y = \sum_{i=1}^m b_i f_i \in l^\infty/c_0$. We claim $\|y\| = 1$. We construct $t \in c_0$ as follows: let $G = \cup_{i,j=1}^m (u_{a_i} \cap u_{a_j}) \subset \mathbb{N}$ and note G is finite by Lemma 7. Set $t_k = -\sum_{i=1}^m b_i \chi_{u_{a_i}}$ if $k \in G$ and $t_k = 0$ otherwise. Now

$$\|y\| = \left\|\sum_{i=1}^m b_i f_i\right\| = \inf\left\{\left\|\sum_{i=1}^m b_i \chi_{u_{a_i}} + z\right\| : z \in c_0\right\}$$

$$\leq \left\|\sum_{i=1}^m b_i \chi_{u_{a_i}} + t\right\| = \sup\left\{\left\|\sum_{i=1}^m (b_i \chi_{u_{a_i}})_k + t_k\right\| : k \in \mathbb{N}\right\}$$

which implies $\|y\| \leq 1$. It is clear that $\|y\| \geq 1$ since $\|\chi_{u_a}\| = 1$ for every $a \in A$. Hence, $\|y\| = 1$.

Now $g(y) = \sum_{i=1}^m [signg(f_i)]g(f_i) = \sum_{i=1}^m |g(f_i)| \geq m/n$ so $\|g\| = \|g\| \|y\| \geq |g(y)| \geq m/n$. Thus, C_n is finite and C is countable.

If $\{g_i\}$ is a countable subset of $(l^\infty/c_0)'$, there exist at most countably many $\pi\chi_{u_a}$ such that $g_i(\pi\chi_{u_a}) \neq 0$ for some i. Thus, there is some $\pi\chi_{u_a}$ such that $g_i(\pi\chi_{u_a}) = 0$ for all i since A is uncountable. Therefore, $\{g_i\}$ does not separate the points of l^∞/c_0 and l^∞/c_0 does not have CTS.

Exercises.

1. Give an example of a projection from c onto c_0 and compute its norm.

2. Show any projection from c onto c_0 has norm greater than 1.

3. Let P_1, P_2 be projections in X. Set $M_i = \mathcal{R}P_i$, $N_i = \ker P_i$. Show $P_1 P_2 = P_2$ iff $M_2 \subset M_1$ iff $N_1 \subset N_2$.

4. Show $P_1 \leq P_2$ iff $P_1 P_2 = P_2 P_1 = P_1$ defines a partial order on the projections in $L(X)$.

5. Show the identity operator on c_0 does not have a continuous linear extension to l^∞ (Recall Exercise 5.4 concerning the extension of continuous linear operators on Hilbert spaces.).

Chapter 12

Schauder Basis

Let X be a normed space.

Definition 12.1. A sequence $\{b_j\} \subset X$ is a Schauder basis for X if each $x \in X$ has a unique representation $x = \sum_{j=1}^{\infty} t_j b_j$, where the series converges in X.

If $\{b_j\}$ is a Schauder basis for X and $x = \sum_{j=1}^{\infty} t_j b_j$, the linear functionals $f_j : X \to \Bbbk$ defined by $f_j(x) = t_j$ are called the *coordinate functionals* relative to $\{b_j\}$. Note $f_i(b_j) = \delta_{ij}$ and $x = \sum_{j=1}^{\infty} f_j(x) b_j$.

Example 12.2. The vectors $\{e^j\}$ form a Schauder basis in c_{00}, c_0, l^p ($1 \leq p < \infty$).

Example 12.3. The space c also has a Schauder basis; namely, $\{e\} \cup \{e^j : j \in \mathbb{N}\}$, where e is the constant sequence of all 1's. For if $t = \{t_j\} \in c$ and $t_0 = \lim t_j$, then $t = t_o e + \sum_{j=1}^{\infty} (t_j - t_0) e^j$.

Note that any normed space with a Schauder basis is separable. In the 1920's Schauder introduced the notion of the Schauder basis and advanced the conjecture that every separable Banach space has a Schauder basis. This was shown not to be the case by P. Enflo in the 1970's when he gave an example of a reflexive Banach space without a Schauder basis ([En]).

Example 12.4. l^∞ does not have a Schauder basis since it is not separable.

The space $C[0, 1]$ also has a Schauder basis which was constructed by Schauder; see [Mo] for a description and other examples.

We now use the Open Mapping Theorem to show that if X is a Banach space with a Schauder basis, then the coordinate functionals relative to the basis are continuous. First, we give an example where the coordinate functionals are not continuous.

Example 12.5. If $\{b_j\}$ is a Schauder basis and $\{f_j\}$ are the coordinate functionals relative to the $\{b_j\}$, then f_1 cannot be continuous if $b_j \to b_1$ for in this case $f_1(b_j) = 0$ for $j > 1$ while $f_1(b_1) = 1$. We construct an example of a Schauder basis with this property in c_{00}. Put $b_1 = e^1$, $b_j = e^1 + \frac{1}{j}e^j$ for $j \geq 2$. Then $b_j \to b_1$. Moreover, $\{b_j\}$ is a Schauder basis for c_{00} for if $t = (t_1, ... t_n, 0, 0, ...) \in c_{00}$, then set $s_1 = t_1 - \sum_{j=1}^{n} j t_j, s_2 = 2t_2, ..., s_n = n t_n$ so $t = \sum_{j=1}^{n} s_j b_j$.

Definition 12.6. A Schauder basis $\{b_j\}$ is monotone if for each $x = \sum_{j=1}^{\infty} t_j b_j$, the sequence $\{\left\|\sum_{j=1}^{n} t_j b_j, \right\|\}_n$ is monotone increasing. Note in this case

$$\left\| \sum_{j=1}^{n} t_j b_j, \right\| \uparrow \|x\|.$$

The Schauder bases in Examples 2 and 3 are monotone; the one in Example 5 is not. We first observe that the coordinate functionals in a space with a monotone basis are always continuous.

Theorem 12.7. *Let $\{b_j\}$ be a monotone Schauder basis for X. Then the coordinate functionals $\{f_j\}$ relative to $\{b_j\}$ are continuous.*

Proof: Fix j. Then

$$(*) \quad |f_j(x)| = \frac{1}{\|b_j\|} \|f_j(x)b_j\| = \frac{1}{\|b_j\|} \left\| \sum_{i=1}^{j} f_i(x)b_i - \sum_{i=1}^{j-1} f_i(x)b_i \right\|$$

$$\leq \frac{1}{\|b_j\|} \left\| \sum_{i=1}^{j} f_i(x)b_i \right\| + \frac{1}{\|b_j\|} \left\| \sum_{i=1}^{j-1} f_i(x)b_i \right\| \leq 2 \|x\| / \|b_j\|$$

by monotonicity.

We next show that a Banach space with a Schauder basis can be renormed so the basis is monotone with respect to the new norm.

Theorem 12.8. *Let X be a Banach space and let $\{b_j\}$ be a Schauder basis for X. Then $\|x\|' = \left\| \sum_{j=1}^{\infty} t_j b_j \right\|' = \sup_n \left\| \sum_{j=1}^{n} t_j b_j \right\|$ defines a complete norm on X which is equivalent to the original norm and under which $\{b_j\}$ is a monotone basis.*

Proof: We first show that $\|\cdot\|'$ defines a complete norm; it is obviously a norm on X and satisfies $\|x\| \leq \|x\|'$. Let $\{x_j\}$ be Cauchy with respect to $\|\cdot\|'$.

From $(*)$ $|f_j(x_n) - f_j(x_m)| \leq \frac{2}{\|b_j\|} \|x_n - x_m\|$ so $\{f_j(x_n)\}_n$ is Cauchy for every j. Set $t_j = \lim_n f_j(x_n)$. We claim that $\sum_{j=1}^{\infty} t_j b_j$ converges in X with respect to $\|\cdot\|$. Let $\epsilon > 0$. Since $\{x_j\}$ is Cauchy with respect to $\|\cdot\|'$, there exists N such that $n, m \geq N$ implies $\left\| \sum_{i=1}^{k} (f_i(x_n) - f_i(x_m)) b_i \right\| \leq \|x_n - x_m\|' < \epsilon$ for all k. Letting $m \to \infty$ gives

$$(\#) \quad \left\| \sum_{i=1}^{k} (f_i(x_n) - t_i) b_i \right\| \leq \epsilon \ for \ n \geq N, k \in \mathbb{N}.$$

Then

$$(\#\#). \ \left\| \sum_{j=1}^{k+p} t_j b_j - \sum_{j=1}^{k} t_j b_j \right\| \leq \left\| \sum_{j=1}^{k+p} t_j b_j - \sum_{j=1}^{k+p} f_j(x_N) b_j \right\| +$$
$$\left\| \sum_{j=1}^{k} t_j b_j - \sum_{j=1}^{k} f_j(x_N) b_j \right\| + \left\| \sum_{j=1}^{k+p} f_j(x_N) b_j - \sum_{j=1}^{k} f_j(x_N) b_j \right\|$$
$$\leq 2\epsilon + \left\| \sum_{j=1}^{k+p} f_j(x_N) b_j - \sum_{j=1}^{k} f_j(x_N) b_j \right\|.$$

Since $\sum_{j=1}^{\infty} f_j(x_N) b_j$ converges , the last term on the right hand side of the inequality in $(\#\#)$ can be made less than ϵ for large k. Thus, the series $\sum_{j=1}^{\infty} t_j b_j$ converges to some $x \in X$ with respect to $\|\cdot\|'$. From $(\#)$ $\|x_n - x\|' \leq \epsilon$ for $n \geq N$. Thus, $\{x_j\}$ converges to x with respect to $\|\cdot\|'$.

It now follows from the Open Mapping Theorem that the norms $\|\cdot\|'$ and $\|\cdot\|$ are equivalent (Proposition 10.14) and $\{b_j\}$ is obviously a monotone basis with respect to $\|\cdot\|'$.

From Theorems 7 and 8, we have

Corollary 12.9. *If X is a Banach space with a Schauder basis $\{b_j\}$, then the coordinate functionals with respect to $\{b_j\}$ are continuous.*

There is a very nice and useful criterion for compactness in a space with a Schauder basis.

Theorem 12.10. *Let X be a Banach space with a Schauder basis $\{b_j\}$ and associated coordinate functionals $\{f_j\}$. A subset $K \subset X$ is relatively compact iff $\lim_n \sum_{j=n}^{\infty} f_j(x) b_j = 0$ uniformly for $x \in K$.*

Proof: Suppose K is relatively compact. By Theorem 8 we may assume that $\{b_j\}$ is a monotone basis. For $n \in \mathbb{N}$ define $R_n, S_n : X \to X$ by $R_n(x) = \sum_{j=n+1}^{\infty} f_j(x) b_j$ and $S_n(x) = \sum_{j=1}^{n} f_j(x) b_j$. Let $\epsilon > 0$. There exists a finite ϵ-net $\{x_1, ..., x_p\}$ for K. Since $\lim_n R_n(x_j) = 0$ for $1 \leq j \leq p$, there exists N such that $\|R_n(x_j)\| < \epsilon$ for $n \geq N$, $1 \leq j \leq p$. If $x \in K$, there exists j such that $\|x - x_j\| < \epsilon$ so if $n \geq N$,

$$\|R_n(x)\| = \|x - S_n(x)\| \leq \|x - x_j\| + \|S_n(x_j) - S_n(x)\| + \|R_n(x_j)\| < 3\epsilon$$

by the monotone property.

Conversely, assume there exists N such that $\|R_n(x)\| < \epsilon$ for $x \in K, n \geq N$. Let $K_N = \{S_N(x) : x \in K\}$. Then $K_N \subset span\{b_1, ..., b_N\}$ and K_N is bounded so K_N is relatively compact in $span\{b_1, ..., b_N\}$ and, therefore, has a finite ϵ-net $\{x_1, ..., x_p\}$. Then $\{x_1, ..., x_p\}$ is a finite 2ϵ-net for K since if $x \in K$, there exists j such that $\|S_N(x) - x_j\| < \epsilon$ and $\|x - x_j\| \leq \|S_N(x) - x_j\| + \|R_N(x)\| < 2\epsilon$.

If X, Y are Banach spaces with Schauder bases $\{x_j\}, \{y_j\}$, respectively, then every operator $A \in L(X, Y)$ has a matrix representation as follows. Let $x \in X$ with $x = \sum_{l=1}^{\infty} t_l x_l$. Each Ax_j has a representation $Ax_j = \sum_{l=1}^{\infty} a_{lj} y_l$. Therefore,

$$y = Ax = \sum_{l=1}^{\infty} t_l Ax_l = \sum_{l=1}^{\infty} t_l \sum_{k=1}^{\infty} a_{kl} y_k$$

and $y \in Y$ so $y = \sum_{l=1}^{\infty} s_l y_l$. Let $\{f_l\}$ be the coordinate functionals associated with $\{y_l\}$. Then

$$s_i = f_i(y) = \sum_{l=1}^{\infty} t_l f_i(\sum_{k=1}^{\infty} a_{kl} y_k) = \sum_{l=1}^{\infty} t_l \sum_{k=1}^{\infty} a_{kl} f_i(y_k) = \sum_{l=1}^{\infty} t_l a_{il}.$$

Hence, A is represented by the infinite matrix $A = [a_{ij}]$ and Ax can be computed by doing the formal matrix multiplication $[a_{ij}][t_j] = [s_i]$.

It is an important problem to give characterizations of the matrices mapping one sequence space into another. As an example of the characterization of a matrix representation as described above, consider a continuous linear operator $A : c_0 \to c_0$. Let $A = [a_{ij}]$ be the matrix representation of A with respect to the Schauder basis $\{e^j\}$ of c_0. Since $\sum_{j=1}^{\infty} a_{ij} t_j$ converges for every $t = \{t_j\} \in c_0$, each row $R_i = \{a_{ij}\}_j \in l^1$ (Exercise 8.1), and since $\lim_i R_i(t) = 0$ for every $t \in c_0$,

$$(*) \quad \sup\{\|R_i\|_1 : i \in \mathbb{N}\} = M < \infty$$

by the Uniform Boundedness Principle. We also have

$$(**) \quad \lim_i Ae^j = \lim_i a_{ij} = 0 \ for \ every \ j.$$

It is also the case that if an infinite matrix $A = [a_{ij}]$ satisfies conditions $(*)$ and $(**)$, then A defines a continuous linear operator from c_0 into

c_0. For, by condition (*), $\left| \sum_{j=1}^{\infty} a_{ij} t_j \right| \leq M \|t\|_{\infty}$ for every $t \in c_0$ so $\|At\|_{\infty} \leq M \|t\|_{\infty}$ and $A : c_0 \to l^{\infty}$ is continuous. Let $\epsilon > 0$. Fix $t \in c_0$. There exists N such that $n \geq N$ implies $|t_n| < \epsilon$. Then

$$(\#) \quad \left| \sum_{j=1}^{\infty} a_{ij} t_j \right| \leq \left| \sum_{j=1}^{N-1} a_{ij} t_j \right| + \left| \sum_{j=N}^{\infty} a_{ij} t_j \right| \leq \left| \sum_{j=1}^{N-1} a_{ij} t_j \right| + M\epsilon$$

and by (**), the first term on the right hand side of (#) can be made less than ϵ for large i.

Thus, conditions (*) and (**) characterize matrices mapping c_0 into itself continuously.

We will give another example of a matrix representation for operators mapping l^1 into l^p later in the next chapter.

Finally, we use the Closed Graph Theorem to establish an automatic continuity result for matrix mappings. Let E be a sequence space with a norm topology. The space E is a *K-space* if the coordinate functionals $\{t_j\} \to t_j$ are continuous from E into the scalar field $\Bbbk$ for every j; the space E is a *BK-space* if E is a K-space and is a Banach space under its norm; the space E is an *AK-space* if E is a K-space and the vectors $\{e^j\}$ form a Schauder basis for E (this terminology is derived from the German words koordinaten and abschnitt). For example, l^p ($1 \leq p < \infty$) and c_0 are AK-BK spaces; l^{∞} and c are BK-spaces but not AK-spaces and c_{00} is an AK-space but not a BK-space.

First, a lemma.

Lemma 12.11. *Suppose E is an AK-BK-space and $a = \{a_j\} \subset \Bbbk$ is such that $a(t) = \sum_{j=1}^{\infty} a_j t_j$ converges for every $t = \{t_j\} \in E$. Then $a \in E'$.*

Proof: For each n the linear functional $a^n(t) = \sum_{j=1}^{n} a_j t_j$ is continuous on E since E is a K-space. Since $a^n(t) \to a(t)$ for every $t \in E$, a is a continuous linear functional by the Banach-Steinhaus Theorem 8.4.

Theorem 12.12. *Let E be an AK-BK-space, F be a BK-space and A an infinite matrix which maps E into F. Then A is continuous.*

Proof: We use the Closed Graph Theorem. Suppose $t^k = \{t_j^k\}_{j=1}^{\infty} \to t = \{t_j\}$ in E and $At^k \to s = \{s_j\}$ in F. We show $At = s$. Let $a^i = \{a_{ij}\}_{j=1}^{\infty}$ be the ith row of A. Since $A : E \to F$, each $a^i \in E'$ by Lemma 11. Therefore, $\lim_k a^i(t^k) = \lim_k \sum_{j=1}^{\infty} a_{ij} t_j^k = a^i(t) = \sum_{j=1}^{\infty} a_{ij} t_j$ for every i. Since F is a K-space, $\lim_k \sum_{j=1}^{\infty} a_{ij} t_j^k = s_i$ for every i. Therefore, $s_i = \sum_{j=1}^{\infty} a_{ij} t_j$ or $s = At$ and the result follows from the Closed Graph Theorem.

It follows from Theorem 12 that any matrix A which maps c_0 into c_0 satisfies conditions (*) and (**).

There is an interesting American Mathematical Monthly article on Schauder basis by R. C. James in [J2].

Exercises.

1. Let $R(\{t_j\}) = (0, t_1, t_2, ...)$ (right shift) and $L(\{t_j\}) = (t_2, t_3, ...)$ (left shift). Give the matrix representations of R^k, L^k as operators from l^p into l^p with respect to $\{e^j\}$ ($1 \le p < \infty$).

2. Let $\{t_j\} \in l^\infty$ and define $T : l^1 \to l^1$ by $Ts = \{t_j s_j\}$. Give the matrix representation of T with respect to $\{e^j\}$.

3. Give a matrix representation for the summation operator $S\{t_j\} = \{\sum_{i=1}^{j} t_i\}$ from l^1 into c.

4. Give a Schauder basis for cs.

5. Let $T \in L(X)$ be invertible. Show $\{b_j\}$ is a Schauder base for X iff $\{Tb_j\}$ is a Schauder base for X.

6. Show the Hilbert cube $C = \{t \in l^2 : |t_j| \le 1/j$ for all $j\}$ is compact in l^2.

Chapter 13

Transpose and Adjoints of Continuous Linear Operators

For any continuous linear operator between normed spaces, there is an associated continuous linear operator acting between the dual spaces, called the transpose operator, which reflects many of the properties of the original operator. We now define and study the properties of the transpose operator.

Let X, Y be normed spaces and $T \in L(X, Y)$. The *transpose* of T, T', is the linear operator from Y' into X' defined by $T'y'(x) = y'(Tx)$, $y' \in Y', x \in X$, i.e., $T'y' = y'T$. We have the following properties of the transpose operator.

Theorem 13.1. (1) $T' \in L(Y', X')$ and $\|T'\| = \|T\|$, (2) if $T, S \in L(X, Y)$, then $(T + S)' = T' + S'$, (3) if $a \in \Bbbk$, then $(aT)' = aT'$. Thus, $T \rightarrow T'$ defines a linear isometry from $L(X, Y)$ into $L(Y', X')$.

Proof: (1): For $y' \in Y'$, $\|T'y'\| \le \|T\| \|y'\|$ so $T' \in L(Y', X')$ and $\|T'\| \le \|T\|$. Using Corollary 7.5, we have
$$\|T'\| = \sup\{\|T'y'\| : \|y'\| = 1\} = \sup\{|T'y'(x)| : \|y'\| = 1, \|x\| = 1\}$$
$$= \sup\{|y'(Tx)| : \|y'\| = 1, \|x\| = 1\} = \sup\{\|Tx\| : \|x\| = 1\} = \|T\|.$$
(2) and (3) are easily checked.

We consider when the mapping $T \rightarrow T'$ from $L(X, Y)$ into $L(Y', X')$ is onto. For this the following observation is useful. Let $T \in L(X, Y)$. It is easily checked that $T''J_X = J_Y T$ or $J_Y^{-1}T''J_X = T$, where J_X is the canonical imbedding of X into X''; if we identify $J_X X$ with X and $J_Y Y$ with Y under the canonical imbeddings, this implies that T'' "extends" T.

Theorem 13.2. The transpose map $T \rightarrow T'$ is onto $L(Y', X')$ iff Y is reflexive (here $X \ne \{0\}$).

Proof: Assume Y is reflexive. Let $A \in L(Y', X')$. Define $T \in L(X, Y)$ by $T = J_Y^{-1}A'J_X$ (note the observation above). Then $T' = A$ since for

$y' \in Y', x \in X,$

$$T'y'(x) = y'(Tx) = y'(J_Y^{-1}A'J_X x) = A'J_X x(y') = J_X x(Ay') = Ay'(x).$$

Therefore, $T' = A$.

Let $y'' \in Y''$. Choose $x_0' \in X', x_0 \in X$ such that $x_0'(x_0) = 1$. Define a continuous linear operator $A : Y' \to X'$ by $Ay' = y''(y')x_0'$. By hypothesis there exists $T \in L(X, Y)$ such that $T' = A$ so

$$y'(Tx_0) = T'y'(x_0) = Ay'(x_0) = y''(y')x_0'(x_0) = y''(y')$$

and $y'' = J_Y T x_0$. Therefore, Y is reflexive.

We use Theorem 2 to give a characterization of the matrices mapping l^1 into l^p for $1 < p < \infty$. Let $1 < p < \infty$ and $\frac{1}{p} + \frac{1}{q} = 1$. Let $T \in L(l^1, l^p)$ and $[t_{ij}]$ be the matrix representation of T with respect to $\{e^j\}$ so

$$\begin{aligned}
\|T\| &= \sup\{\|Tx\|_p : \|x\|_1 = 1\} = \sup\{|y'Tx| : \|x\|_1 = 1, \|y'\|_q = 1\} \\
&= \sup\{\|T'y'\|_\infty : \|y'\|_q = 1\} = \sup\left|T'y'(e^j)\right| : \|y'\|_q = 1, j \in \mathbb{N}\} \\
&= \sup\{\left\|Te^j\right\|_p : j \in \mathbb{N}\} = \sup\{\left\|\textstyle\sum_{i=1}^\infty t_{ij}e^i\right\|_p : j \in \mathbb{N}\} \\
&= \sup\{(\textstyle\sum_{i=1}^\infty |t_{ij}|^p)^{1/p} : j \in \mathbb{N}\} < \infty.
\end{aligned}$$

We show conversely, that if a matrix $[t_{ij}]$ satisfies

$$M = \sup\{(\sum_{i=1}^\infty |t_{ij}|^p)^{1/p} : j \in \mathbb{N}\} < \infty,$$

then it defines a continuous linear operator $T \in L(l^1, l^p)$. For each j ,the column $(t_{1j}, t_{2j}, ...) \in l^p$ induces a continuous linear functional f_j on l^q with $\|f_j\| = (\sum_{i=1}^\infty |t_{ij}|^p)^{1/p} \le M$. Define $S : l^q \to l^\infty$ by $Sx = \{f_j(x)\}$. Note $S \in L(l^q, l^\infty)$ since $\|Sx\|_\infty = \sup\{|f_j(x)| : j\} \le M \|x\|_q$. By Theorem 2 there exists $T \in L(l^1, l^p)$ such that $T' = S$. We claim the matrix representation of T is $[t_{ij}]$. For $i, j \in \mathbb{N}$, we have

$$e^i(Te^j) = Se^i(e^j) = (\sum_{k=1}^\infty f_k(e^i)e^k)(e^j) = \sum_{k=1}^\infty f_k(e^i)e^k(e^j) = f_j(e^i) = t_{ij}$$

so $Te^j = \sum_{i=1}^\infty t_{ij}e^i$.

Note that the transpose of the identity operator is the identity operator. We also have the following observation.

Proposition 13.3. *Let X, Y, Z be normed spaces with $T \in L(X, Y), S \in L(Y, Z)$. Then $(ST)' = T'S'$.*

Example 13.4. If $T \in L(l^2)$, then T has a matrix representation $T = [t_{ij}]$ with respect to the basis $\{e^j\}$. Then $T' \in L(l^2)$ also has a matrix representation $T' = [s_{ij}]$ with respect to $\{e^j\}$. Since $T'e^j(e^i) = e^j(Te^i) = t_{ji}$, T' has $[t_{ji}] = [s_{ij}]$ as its matrix representation, i.e., the matrix representation of T' is given by the formal adjoint matrix of T.

We now consider the relationship between an operator and its transpose.

Theorem 13.5. *Let X, Y be normed spaces and $T \in L(X, Y)$. T has a continuous inverse with domain Y iff T' has a continuous inverse with domain X'. Moreover, $(T^{-1})' = (T')^{-1}$.*

Proof: If $T^{-1} \in L(Y, X)$, then $(TT^{-1})' = (I_Y)' = I_{Y'} = (T^{-1})'T'$ and $(T^{-1}T)' = T'(T^{-1})' = (I_X)' = I_{X'}$ so $(T')^{-1}$ exists and $(T')^{-1} = (T^{-1})'$ so $(T')^{-1} \in L(X', Y')$.

If $(T')^{-1} \in L(X', Y')$, then by the part above $(T'')^{-1}$ exists and is in $L(X'', Y'')$. Therefore, T'' is a linear homeomorphism and extends T by the observation above. Hence, T is one-one and TX is closed. Thus, we must show that $TX = Y$. Suppose there exists $y \in Y \backslash TX$. By Corollary 7.4 there exists $y' \in Y'$ such that $y'(y) \neq 0$ and $y'(TX) = 0$, i.e., $y'T = T'y' = 0$. But then T' is not one-one since $y' \neq 0$. This contradiction shows $TX = Y$.

The proof of Theorem 5 establishes

Corollary 13.6. *If $T \in L(X, Y)$, then*

$$\overline{TX} = \{y \in Y : y'(y) = 0 \text{ for every } y' \text{ such that } T'y' = 0\}.$$

In what follows X, Y are normed spaces and $T \in L(X, Y)$. The range of T is denoted by $\mathcal{R}T$ and the kernel by $\mathcal{N}T$ (nullity of T). The domain of an operator S is denoted by $\mathcal{D}S$.

Theorem 13.7. *If T' has a continuous inverse (on $\mathcal{R}T'$), then $\mathcal{R}T'$ is closed.*

Proof: Suppose $T'y'_j \to x' \in X'$. By Theorem 2.11 there exists $m > 0$ such that $\left\| T'y'_j - T'y'_k \right\| \geq m \left\| y'_j - y'_k \right\|$ which implies $\{y'_j\}$ is Cauchy. Let $y'_j \to y' \in Y'$. Since T' is continuous, $T'y'_j \to T'y' = x'$. Therefore, $x' \in \mathcal{R}T'$ and $\mathcal{R}T'$ is closed.

Theorem 13.8. $\mathcal{R}T' = X'$ *iff T has a bounded inverse.*

Proof: Suppose T has a bounded inverse. For $x' \in X'$, $x'T^{-1}$ is a continuous linear functional on $\mathcal{R}T = \mathcal{D}(T^{-1})$. By the Hahn-Banach Theorem $x'T^{-1}$ has a continuous linear extension to Y', say y'. Then $y'(Tx) = x'T^{-1}(Tx) = x'(x)$ for $x \in X$ so $x' = T'y'$ and $\mathcal{R}T' = X'$.

Suppose $\mathcal{R}T' = X'$ and T does not have a continuous inverse. By Theorem 2.11 there exists $\{x_j\} \subset X$ such that $\|Tx_j\| / \|x_j\| \to 0$. Let

$$a_j = \max\{(\|Tx_j\| / \|x_j\|)^{1/2}, 1/\sqrt{j}\}$$

and $u_j = x_j/a_j \|x_j\|$. Then $\|u_j\| \to \infty$ and

$$\|Tu_j\| = \|Tx_j\| / a_j \|x_j\| \le (\|Tx_j\| / \|x_j\|)^{1/2} \to 0.$$

Thus, $y'(Tu_j) \to 0$ for every $y' \in Y'$, i.e., $T'y'(u_j) \to 0$ or since $\mathcal{R}T' = X'$, $x'(u_j) \to 0$ for every $x' \in X'$. By Proposition 9.4, $\{\|u_j\|\}$ is bounded. But, $\|u_j\| \to \infty$.

Theorem 13.9. *Suppose Y is a Banach space and $\mathcal{R}T = Y$. Then T' has a continuous inverse.*

Proof: If T' does not have a continuous inverse, then as in the proof of Theorem 8 there exists a sequence $\{y_j'\} \subset Y'$ such that $\|y_j'\| \to \infty$ while $\|T'y_j'\| \to 0$. Therefore, for every $x \in X$, $T'y_j'(x) = y_j'T(x) \to 0$. Since $\mathcal{R}T = Y$, $y_j'(y) \to 0$ for every $y \in Y$. Since Y is complete, by the Uniform Boundedness Principle, $\{\|y_j'\|\}$ is bounded giving a contradiction.

Definition 13.10. If $M \subset X$, the annihilator of M in X' is

$$M^0 = \{x' \in X' : x'(x) = 0 \text{ for all } x \in M\}.$$

Note that M^0 is a closed subspace of X, and if M is a subspace, then $M^0 = (\overline{M})^0$.

Definition 13.11. If $N \subset X'$, the annihilator of N in X is

$$N_0 = \{x \in X : x'(x) = 0 \text{ for all } x' \in N\}.$$

Note that N_0 is a closed subspace of X, and if N is a subspace, then $N_0 = (\overline{N})_0$.

Proposition 13.12. *If M is a subspace of X, then $(M^0)_0 = \overline{M}$.*

Proof: Let $m \in M$. Then for $f \in M^0$, $f(m) = 0$ which implies $m \in (M^0)_0$. Since $(M^0)_0$ is closed, we have $\overline{M} \subset (M^0)_0$.

Let $x \in (M^0)_0$. Then $f(x) = 0$ for $f \in M^0$. If $x \notin \overline{M}$, then by Corollary 7.4 there exists $f \in X'$ such that $f(\overline{M}) = 0$ but $f(x) \neq 0$. Thus, $x \in \overline{M}$ and $(M^0)_0 \subset \overline{M}$.

The dual of Proposition 12 is given by

Proposition 13.13. *If N is a subspace of X', then $\overline{N} \subset (N_0)^0$.*

The containment in Proposition 13 can be proper. For example, consider $X = c$; the dual of c is l^1 under the pairing $s(t) = \sum_{k=1}^{\infty} s_k t_{k-1}$, where $s \in l^1, t \in c$ and $t_0 = \lim_k t_k$. Let F be the subspace of l^1 consisting of all $s = \{s_k\}$ with $s_1 = 0$. Then F is a proper closed subspace of l^1 and $F_0 = \{0\}$ since F contains the coordinate functionals $s \to s_k$ for all k. Then $(F_0)^0 = l^1$. However, if X is reflexive, equality holds in Proposition 13 (see Exercise 7).

We have the following relationships between ranges, kernels and annihilators.

Proposition 13.14. *(1) $(\overline{\mathcal{R}T})^0 = \mathcal{N}(T')$, (2) $\overline{\mathcal{R}T} = \mathcal{N}(T')_0$, (3) $\overline{\mathcal{R}T} = Y$ iff T' is one-one.*

Proof: (1): Let $y' \in \mathcal{N}(T')$ so $T'y' = 0, i.e., T'y'(x) = y'Tx = 0$ for all $x \in X$. Hence, $y' \in \mathcal{R}T^0$ and $\mathcal{N}(T') \subset \mathcal{R}T^0$ and since $\mathcal{R}T^0 = (\overline{\mathcal{R}T})^0$, $\mathcal{N}(T') \subset (\overline{\mathcal{R}T})^0$.

Let $y' \in (\overline{\mathcal{R}T})^0 = \mathcal{R}T^0$. Then $y'(Tx) = T'y'(x) = 0$ so $y' \in \mathcal{N}(T')$. Therefore, $(\overline{\mathcal{R}T})^0 \subset \mathcal{N}(T')$.

(2): $(\overline{\mathcal{R}T})^0 = \mathcal{N}(T')$ implies by Proposition 12, $((\overline{\mathcal{R}T})^0)_0 = \mathcal{N}(T')_0 = \overline{\mathcal{R}T}$.

(3): $\overline{\mathcal{R}T} = Y$ iff $\mathcal{N}(T')_0 = Y$ by (2) iff $\mathcal{N}(T') = \{0\}$ by Theorem 7.2 iff T' is one-one.

The dual of Proposition 14 is given by

Proposition 13.15. *(1) $(\overline{\mathcal{R}T'})_0 = \mathcal{N}(T)$, (2) $\overline{\mathcal{R}T'} \subset \mathcal{N}(T)^0$, (3) If $\overline{\mathcal{R}T'} = X'$, then T is one-one.*

Theorem 13.16. $\overline{\mathcal{R}T} = Y$ *and T has a continuous inverse iff $\mathcal{R}T' = X'$ and T' has a continuous inverse.*

Proof: By Proposition 14.(3) and Theorem 8, $\overline{\mathcal{R}T} = Y$ and T has a continuous inverse iff T' is one-one and $\mathcal{R}T' = X'$ iff T' is a homeomorphism by the Open Mapping Theorem 10.12.

Lemma 13.17. *Let X, Y be Banach spaces. Suppose T is onto. Then there exists $c > 0$ such that for every $y \in Y$ there exists $x \in X$ with $Tx = y$ and $\|x\| \le c\|y\| = c\|Tx\|$.*

Proof: Let π be the quotient map from X onto $X/\mathcal{N}(T)$ and let $\widehat{T}$ be the induced map from $X/\mathcal{N}(T)$ onto Y. Then by the Open Mapping Theorem $\widehat{T}$ has a continuous inverse. Let $c = \left\|(\widehat{T})^{-1}\right\| + 1$. Then for $y \ne 0, y \in Y$,

$$c\|y\| > \left\|(\widehat{T})^{-1}\right\|\|y\| \ge \left\|(\widehat{T})^{-1}y\right\|$$

so the existence of x follows from the definition of the quotient norm.

Theorem 13.18. *Let X, Y be Banach spaces and $T \in L(X, Y)$. If $\mathcal{R}T$ is closed, then $\mathcal{R}T'$ is closed.*

Proof: We claim that $\mathcal{R}T' = \mathcal{N}(T)^0$ (see Proposition 15(2)); this will establish the result. Let $x' \in \mathcal{N}T^0$. Define a linear functional f_1 on $\mathcal{R}T$ $= \overline{\mathcal{R}T}$ by $f_1 Tx = x'(x)$; f_1 is well-defined since if $Tx_1 = Tx_2$, then $x_1 - x_2 \in \mathcal{N}T$ and $x'(x_1) = x'(x_2) = f_1(Tx_1) = f_1(Tx_2)$. f_1 is clearly linear and we claim that f_1 is continuous. For if $y_j \in \mathcal{R}T$ and $\|y_j\| \to 0$, choose $x_j \in X$ such that $Tx_j = y_j$ as in Lemma 17. (Here we are considering T as a linear operator from X onto $\mathcal{R}T$.) By Lemma 17 $\|x_j\| \to 0$ so $f_1(y_j) = x'(x_j) \to 0$ since x' is continuous. Extend f_1 to a continuous linear functional $y' \in Y'$ by the Hahn-Banach Theorem. Then $y'(Tx) = f_1(Tx) = x'(x)$ which implies $x' = T'y'$ so $x' \in \mathcal{R}T'$. Hence, $\mathcal{N}(T)^0 \subset \mathcal{R}T'$ and the reverse inclusion always holds (Proposition 15).

The converse of Theorem 18 holds but is much more difficult to establish; see [Y] VII.5. Theorem 18 along with its converse is often referred to as Banach's Closed Range Theorem.

We now consider the transpose of projection operators.

Theorem 13.19. *Let X be a Banach space and $P : X \to X$ a projection. Then $P' : X' \to X'$ is a projection with $\mathcal{R}P' = (\ker P)^0$ and $\ker P' = (\mathcal{R}P)^0$.*

Proof: By Proposition 3, $(P')^2 = (P^2)' = P'$ so P' is a projection.

Let $X = M \oplus N$, where M is the range of P and N is the kernel of P. By Proposition 14.(1), $\ker P' = (\mathcal{R}P)^0 = M^0$. By the same result $\mathcal{R}P' = \ker(I - P') = (\mathcal{R}(I - P))^0 = N^0$.

Adjoints in Hilbert space:

If H is a Hilbert space and $T \in L(H)$, then T', the transpose of T, is a linear, continuous operator on H'. But, by the Riesz Representation Theorem 5.19, we may "identify" H and H' and then "T'" can be viewed as a continuous linear operator on H. We refer to this map as the *adjoint* of T and denote it by T^*. We now give the formal definition.

Let $\Psi = \Psi_H$ be the conjugate linear isometry given by the Riesz Representation Theorem, $\Psi h(h_1) = h_1 \cdot h$. Define $T^* \in L(H)$ by $T^* = \Psi^{-1} T' \Psi$. We call T^* the *adjoint* of T. Thus, for $x, y \in H$, the adjoint operator T^* is characterized by the condition

$$Ty \cdot x = y \cdot T^* x$$

since $T' \Psi x(y) = \Psi(x)(Ty) = Ty \cdot x$ and $T' \Psi(x)(y) = \Psi T^* x(y) = y \cdot T^* x$.

From the previously established properties of the transpose operator, we have the following properties of the adjoint operator.

Proposition 13.20. *Let $T, S \in L(H)$. Then (1) $(T + S)^* = T^* + S^*$, (2) $(TS)^* = S^* T^*$, (3) $(aT)^* = \overline{a} T^*$ for $a \in \Bbbk$, (4) $I = I^*$, (5) $T^{**} = T$, (6) $\|T\| = \|T^*\|$, (7) if either T^{-1} or $(T^*)^{-1}$ exists and is in $L(H)$, the other exists and $(T^{-1})^* = (T^*)^{-1}$.*

Example 13.21. Suppose $T \in L(l^2)$, where we consider complex scalars. Let $[t_{ij}]$ be the matrix representation of T with respect to $\{e^j\}$. Then $T^* \in L(l^2)$. Suppose $T^* e^j = \sum_{k=1}^{\infty} b_{kj} e^k$ so

$$T^* e^j \cdot e^i = e^j \cdot T e^i = b_{ji} = e^j \cdot \sum_{k=1}^{\infty} t_{ik} e^k = \overline{t_{ij}}.$$

Thus, T^* is represented by the matrix $[\overline{t_{ji}}]$. Note the difference between the representations for the transpose and adjoint operators, i.e., the appearance of the conjugates.

For later use we establish an important relationship for the adjoint.

Proposition 13.22. *Let $T \in L(H)$. Then $\|T^* T\| = \|T\|^2 = \|TT^*\|$.*

Proof: $\|T^* T\| \leq \|T\| \|T^*\| = \|T\|^2$ by (6) of Proposition 20.
$\|T\|^2 = (\sup\{\|Tx\| : \|x\| \leq 1\})^2 = \sup\{\|Tx\|^2 : \|x\| \leq 1\}$
$= \sup\{Tx \cdot Tx : \|x\| \leq 1\} = \sup\{T^* Tx \cdot x : \|x\| \leq 1\}$
$\leq \sup\{\|T^* Tx\| : \|x\| \leq 1\} = \|T^* T\|$.
The other equality is obtained in a similar manner using $\|T^*\|^2 = \|T\|^2$.

Exercises.

1. Let $1 \leq p < \infty$ and L (R) be the left (right) shift on l^p. Compute the transpose of L (R).

2. Let X be a Banach space. If $T \in L(X, Y)$ has a continuous inverse, show T has a closed range.

3. Let X, Y be Banach spaces and $T \in L(X, Y)$. Show that if T is one-one and has closed range, then T' is onto.

4. Let X, Y be Banach spaces and $T \in L(X, Y)$. If T is one-one and has closed range, show T'' is one-one.

5. Show if T'' is one-one, then T is one-one but not conversely. (Hint: Consider $T \in L(c_0)$, $T\{t_j\} = \{t_{j+1} - t_j\}$.)

6. Use shift operators to show that the containment in Proposition 15.(2) can be proper.

7. Show equality holds in Proposition 13 if X is reflexive.

8. If X is a Banach space, show there is a projection of norm 1 from X''' onto X'.

9. Use Exercise 8 to show that c_0 does not have a *predual*, i.e., there is no Banach space X with $X' = c_0$.

10. Let X be a Banach space and $T \in L(X)$. What is the transpose of e^T? (Recall Exercise 2.14.)

11. If X, Y are Banach spaces with Schauder bases and $T \in L(X, Y)$ is represented by a matrix as in Chapter 12 , what is the matrix representation of T' (if any)?

Chapter 14

Compact Operators

In this chapter we study a class of operators called compact operators. These operators have important applications to both integral and differential equations which will be given in later chapters.

Let X, Y be normed spaces.

Definition 14.1. A linear map $T : X \to Y$ is compact (precompact) if T carries bounded sets into relatively compact (precompact) sets.

Remark 14.2. Note that any compact (precompact) operator is bounded and, therefore, continuous.

As an elementary example note that any linear operator with a finite dimensional range is compact. Also, the identity operator on a normed space X is compact iff X is finite dimensional (Corollary 4.9).

Let $K(X,Y)$ $(PC(X,Y))$ be the space of all compact (precompact) operators from X into Y. Note that if Y is complete, then $K(X,Y) = PC(X,Y)$.

Proposition 14.3. $K(X,Y)$ $(PC(X,Y))$ is a linear subspace of $L(X,Y)$.

Proof: Let $T_1, T_2 \in K(X,Y)$. Let $B \subset X$ be bounded. Then $T_i B$ is relatively compact so $(T_1 + T_2)B$ is relatively compact so $T_1 + T_2 \in K(X,Y)$. Similarly, $aT_1 \in K(X,Y)$ for $a \in \Bbbk$.

The precompact case is similar.

Proposition 14.4. Let Z be a normed space, $T \in L(X,Y)$ and $S \in L(Y,Z)$. If T is compact (precompact), then ST is compact (precompact); if S is compact (precompact), then ST is compact (precompact).

Theorem 14.5. $PC(X,Y)$ is a closed subspace of $L(X,Y)$.

Proof: Let T_j be precompact for $j \in \mathbb{N}$ and $T \in L(X, Y)$ with $\|T_j - T\| \to 0$. Let $\epsilon > 0$. There exists N such that $\|T_N - T\| < \epsilon/3$. Since T_N is precompact, there exist $x_1, ..., x_n \in X$ of norm 1 such that $\{T_N x_i : i = 1, ..., n\}$ is an $\epsilon/3$ net for $\{T_N x : \|x\| \le 1\}$. If $x \in X$, $\|x\| \le 1$, there exists j such that $\|T_N x_j - T_N x\| < \epsilon/3$. Then

$$\|Tx - Tx_j\| \le \|Tx - T_N x\| + \|T_N x - T_N x_j\| + \|T_N x_j - Tx_j\| < \epsilon$$

so $\{T x_i : i = 1, ..., n\}$ is an ϵ-net for $\{Tx : \|x\| \le 1\}$.

Corollary 14.6. *If Y is complete, $K(X, Y)$ is a closed subspace of $L(X, Y)$.*

Example 14.7. Completeness in Corollary 6 is important. Define $T : c_0 \to l^2$ by $T(\{t_i\}) = \{t_i/i\}$. Then $T \in L(c_0, l^2)$. If $s^k = \sum_{j=1}^{k} e^j$, then $\|s^k\|_\infty = 1$ and $T s^k = \sum_{j=1}^{k} e^j/j \to y = \sum_{j=1}^{\infty} e^j/j$ in l^2. Now consider T as a continuous linear operator from c_0 into the range of T, Y_0. Then T is not compact since $\{T s^k\}$ has no convergent subsequence in Y_0 since $y \notin Y_0$. Define $T_k \in K(c_0, Y_0)$ by $T_k(\{t_i\}) = \sum_{j=1}^{k} (t_j/j) e^j$. Now $\|T_k - T\| \to 0$ since

$$\sup\{\|(T_k - T)(\{t_j\})\| : \|\{t_j\}\|_\infty \le 1\} \le \left(\sum_{j=k+1}^{\infty} 1/j^2 \right)^{1/2} \to 0.$$

Remark 14.8. Note that T also furnishes an example of a precompact operator which is not compact.

Corollary 14.9. *If X is a Banach space, then $K(X, X)$ is a closed 2-side ideal in $L(X, X)$.*

There are Banach spaces for which $K(X, X)$ is the only closed 2-sided ideal in $L(X, X)$ (see Goldberg ([Go] p. 84)).

Proposition 14.10. *If $T \in K(X, Y)$, then T has separable range.*

Proof: If $S_n = \{x : \|x\| \le n\}$, then the range of T is $\cup_{n=1}^{\infty} T S_n$ so it suffices to show that each $T S_n$ is separable. Each $T S_n$ is relatively compact and, therefore, has a finite $1/k$-net for every $k \in \mathbb{N}$ so $T S_n$ is separable.

We now give some examples of compact operators.

Example 14.11. Let $k \in C([a, b] \times [a, b])$ and let $K : C[a, b] \to C[a, b]$ be the integral operator induced by k, $Kf(t) = \int_a^b k(t, s) f(s) ds$. We show that K is compact. Let $\{f_j\}$ be bounded in $C[a, b]$. To show that $\{Kf_j\}$ is relatively compact, it suffices by the Arzela-Ascoli Theorem to show that $\{Kf_j\}$ is bounded and equicontinuous ([DeS]26.11). $\{Kf_j\}$ is bounded since K is continuous (Example 2.7). Let $\epsilon > 0$. There exists $\delta > 0$ such that $|k(t,s) - k(t',s')| < \epsilon$ for $t, t', s, s' \in [a, b]$ and $|t - t'| < \delta, |s - s'| < \delta$. Thus, for $t, t' \in [a, b], |t - t'| < \delta$, we have $|Kf_j(t) - Kf_j(t')| \leq \int_a^b |k(t,s) - k(t',s)| \, |f_j(s)| \, ds \leq \|f_j\|_\infty \epsilon(b - a)$. Hence, $\{Kf_j\}$ is equicontinuous.

Example 14.12. The Volterra operator (Example 2.6) $V : C[0,1] \rightarrow C[0,1]$ defined by $Vf(t) = \int_0^t f(s)ds$ is compact (Exercise 1).

Example 14.13. Following Theorem 13.2 we gave a matrix characterization of continuous linear operators from l^1 into l^p for $1 < p < \infty$. We now give a similar characterization of the compact operators from l^1 into l^p for $1 < p < \infty$. If $T \in L(l^1, l^p)$ has the matrix representation $[t_{ij}]$, then T is compact iff

$$(*) \quad \lim_n (\sum_{i=n}^\infty |t_{ij}|^p)^{1/p} = 0 \text{ uniformly for } j \in \mathbb{N}.$$

Suppose that T is compact. Then $\{Te^j = (t_{1j}, t_{2j}, ...) : j \in \mathbb{N}\}$ is relatively compact so by Theorem 12.10, $\lim_n \left\| \sum_{i=n}^\infty t_{ij}e^i \right\|_p = 0$ uniformly for $j \in \mathbb{N}$, i.e., $(*)$ holds. Conversely, suppose that $(*)$ is satisfied. Define $T_n \in K(l^1, l^p)$ by $T_n(\{s_j\}) = \sum_{i=1}^n (\sum_{j=1}^\infty t_{ij}s_j)e^i$. Then $\|T_n - T\| = \sup_j (\sum_{i=n+1}^\infty |t_{ij}|^p)^{1/p} \rightarrow 0$ by Theorem 12.10 and $(*)$. Hence, T is compact by Corollary 6.

The compact operator in Example 13 is the limit of a sequence of operators with finite dimensional range. If the range of a compact operator has a Schauder basis, then every compact operator is the limit of a sequence of operators with finite dimensional range.

Theorem 14.14. *Let Y be a Banach space with a Schauder basis. If $T \in K(X, Y)$, then there exists a sequence of operators, $\{T_j\}$, with finite dimensional range such that $\|T_j - T\| \rightarrow 0$.*

Proof: Let $\{b_j\}$ be a Schauder basis for Y with coordinate functionals $\{f_j\}$ and let S be the closed unit ball of X. Define $T_j : X \rightarrow Y$ by $T_j x = \sum_{i=1}^j f_i(Tx)b_i$. Then T_j has finite dimensional range and so is compact. Let $\epsilon > 0$. Since TS is relatively compact, there exists N such that $n \geq N$ implies $\|\sum_{i=n}^\infty f_i(Tx)b_i\| < \epsilon$ for all $x \in S$ (Theorem 10.12). Thus, $\|T_n x - Tx\| < \epsilon$ for $x \in S, n \geq N$ or $\|T_n - T\| \leq \epsilon$ for $n \geq N$.

A Banach space X has the *approximation property*. if every compact operator $T \in K(X,X)$ is the norm limit of a sequence of operators with finite dimensional range. From Theorem 14, it follows that Banach spaces with a Schauder basis have the approximation property. It was an open question as to whether every Banach space has the approximation property. However, Enflo has given an example of a separable Banach space without the approximation property ([En]).

We next consider the transpose of compact operators.

Theorem 14.15. *(Schauder) Let $T \in L(X,Y)$. Then T is precompact iff T' is compact.*

Proof: Let $T \in PC(X,Y)$ and $\epsilon > 0$. We show that T' is precompact and then T' is compact since Y' is complete. There exists $x_1, ..., x_k$ in $S = \{x : \|x\| \leq 1\}$ such that $Tx_1, ..., Tx_k$ is an $\epsilon/3$-net for TS. Define $A : Y' \to \Bbbk^k$ by $Ay' = (y'Tx_1, ..., y'Tx_k)$. Then A is compact so there exist $y'_1, ..., y'_m$ in $S' = \{y' : \|y'\| \leq 1\}$ such that $Ay'_1, ..., y'Ay'_m$ is an $\epsilon/3$-net for AS'. If $y' \in S'$, there exist i such that $\|Ay' - Ay'_i\| < \epsilon/3$ and, hence, $|y'(Tx_j) - y'_i(Tx_j)| < \epsilon/3$ for $j = 1, ..., k$. Thus, if $x \in S$, there exists j such that $\|Tx_j - Tx\| < \epsilon/3$ and

$$|T'y'(x) - T'y'_i(x)| \leq |y'(Tx - Tx_j)| + |(y' - y'_i)(Tx_j)| + |y'_i(Tx_j - Tx)| < \epsilon$$

so $T'y'_1, ..., T'y'_m$ is an ϵ-net for $T'S'$.

If T' is compact, then T'' is compact by what was established above. If J_X (J_Y) is the canonical imbedding of X (Y) into X'' (Y''), then $T''J_X = J_Y T$ is compact. Hence, T is precompact since J_Y has bounded inverse.

Corollary 14.16. *If Y is a Banach space, then $T \in L(X,Y)$ is compact iff T' is compact.*

The operator in Example 7 shows that completeness is important in Corollary 16.

We next consider some continuity properties for compact operators.

Theorem 14.17. *Let $T \in K(X,Y)$. Then T carries weakly convergent sequences into norm convergent sequences.*

Proof: Suppose $\{x_j\}$ converges weakly to 0, but $\{Tx_j\}$ isn't norm convergent to 0. Then there exist $\epsilon > 0$ and a subsequence $\{x_{n_j}\}$ such that $\|Tx_{n_j}\| > \epsilon$ for every j. Since $\{x_j\}$ is weakly bounded, $\{x_j\}$ is norm bounded (Proposition 9.4) so $\{Tx_{n_j}\}$ has a subsequence which is norm convergent to some $y \subset Y$; for convenience, assume $\|Tx_{n_j} - y\| \to 0$. But, T is weakly sequentially continuous (Proposition 9.5) so $Tx_{n_j} \to 0$ weakly and $y = 0$ which gives the desired contradiction.

Operators with the property in Theorem 17 are called *completely continuous* and were introduced by Hilbert for operators in l^2 where it is equivalent to an operator being compact (see Proposition 18). The definition of compact operators which we now use is due to F. Riesz. The converse of Theorem 17 is false in general; consider the identity operator on l^1 (Corollary 9.15). We do have a partial converse.

Proposition 14.18. *Let X be reflexive. Then $T \in L(X, Y)$ is compact iff T is completely continuous.*

Proof: Assume that T is completely continuous. Let $\{x_j\}$ be bounded in X. Then $\{x_j\}$ has a subsequence $\{x_{n_j}\}$ which is weakly convergent to some $x \in X$ (Theorem 9.6). Then $\|Tx_{n_j} - Tx\| \to 0$ so $\{Tx_j\}$ is relatively compact and T is compact.

Exercises.

1. (General Volterra Operator) Let $k : [0, 1] \times [0, 1] \to \Bbbk$ be continuous and define $Kf(t) = \int_0^t k(t, s) f(s) ds$ for $f \in C[0, 1]$. Show K is compact. See Example 12.

2. Let X , Y be Banach spaces. Suppose there exists a compact operator $T : Y \to X$ which is onto. Show X is finite dimensional.

3. Let X, Y be Banach spaces. If $T \in K(X, Y)$ has closed range, show the range of T is finite dimensional.

4. Let H be a Hilbert space and $\{f_j\}$ be an orthonormal set. If $\{t_j\} \in c_0$, show $Tx = \sum_{j=1}^{\infty} t_j (f_j \cdot x) f_j$ defines a compact operator on H.

5. Show that if $T \in L(X, Y)$ is completely continuous, then T carries weak Cauchy sequences into norm Cauchy sequences.

6. Suppose X' is separable and Y is a Banach space. Show $T \in L(X, Y)$ is completely continuous iff T is compact. [Hint: Use Exercises 5 and 9.4.]

7. Give an example of a completely continuous operator whose transpose is not completely continuous.

8. Let H be a Hilbert space and $T \in K(H, H)$. If $\{f_j\}$ is orthonormal, show $\|Tf_j\| \to 0$.

9. Suppose $g \in C[0, 1]$ and $g \neq 0$. Define $G : C[0, 1] \to C[0, 1]$ by $Gf = gf$. Show G is linear and continuous but not compact. [Hint: g

is bounded away from 0 on some interval J; let $\{f_j\}$ be an orthonormal subset in $C(J)$ and extend it to an orthonormal subset $\{g_j\}$ of $C[0,1]$ and consider $\{Gg_j\}$.]

10. If $\sum_{i,j} |a_{ij}|^2 < \infty$, show $A = [a_{ij}]$ defines a compact operator on l^2.

11. Show a projection $P \in L(X)$ is compact iff the range of P is finite dimensional.

12. Let X, Y be Banach spaces. Show that if $T \in L(X, Y)$ has a closed, infinite dimensional range, then T cannot be compact.

13. Show the summing operator in Example 2.9 is not compact.

Chapter 15

The Fredholm Alternative

The Fredholm Alternative from integral equations concerns the possibilities for solving the Fredholm equation

$$f(t) - \lambda \int_a^b k(t,s)f(s)ds = g(t)$$

for the unknown function f. If the kernel k is continuous and K is the integral operator induced by k, this equation can be written in operator form as $(I - \lambda K)f = g$, where $f, g \in C[a,b]$. In analogy with the situation for finite matrices, the Fredholm Alternative states that the non-homogeneous equation $(I - K)f = g$ has a solution for every g iff the homogeneous equation $(I - K)f = 0$ has only the trivial solution.

In this chapter we establish an abstract version of the Fredholm Alternative for compact operators and give applications of the abstract result to integral equations in later chapters.

Throughout this chapter, let X be a complex Banach space and $T \in K(X,X)$. Let $\lambda \in \mathbb{C}, \lambda \neq 0$. We write $\lambda - T$ for $\lambda I - T$, where I is the identity operator on X.

Theorem 15.1. *If $\lambda - T$ is onto, then $\lambda - T$ is one-one.*

Proof: Suppose there exists $x_1 \neq 0$ such that $Tx_1 = \lambda x_1$. Set $S = \lambda - T$ so $Sx_1 = 0$. Now

$$\ker S \subset \ker S^2 \subset \dots \subset \ker S^n \subset \dots$$

and each of these subspaces is closed. We claim the containments are proper. Since S is onto, there exists x_2 such that $Sx_2 = x_1$ and similarly there exists x_3 such that $Sx_3 = x_2$, etc. Thus, we have a sequence $\{x_n\}$ with $Sx_{n+1} = x_n$ and $x_n \neq 0$ for every n. Now $x_n \in \ker S^n$ since

$$S^n x_n = S^{n-1}(Sx_n) = S^{n-1}x_{n-1} = \dots = Sx_1 = 0.$$

But, $x_n \notin \ker S^{n-1}$ since $S^{n-1}x_n = Sx_2 = x_1 \neq 0$.

By Riesz's Lemma (Lemma 4.6), for each n there exists $y_{n+1} \in \ker S^{n+1}$ such that $\|y_{n+1}\| = 1$ and $\|y_{n+1} - x\| \geq 1/2$ for $x \in \ker S^n$. For $n > m$,

$$\|Ty_n - Ty_m\| = \|\lambda y_n - (\lambda y_m - Sy_n - Sy_m)\|$$
$$= |\lambda| \, \|y_n - (y_m - S(y_n/\lambda) - S(y_m/\lambda))\| \geq |\lambda|/2$$

since $(y_m - S(y_n/\lambda) - S(y_m/\lambda)) \in \ker S^{n-1}$. Thus, $\{y_n\}$ is a bounded sequence such that $\{Ty_n\}$ has no convergent subsequence. This contradicts the compactness of T.

Theorem 15.2. *The range of $\lambda - T$ is closed.*

Proof: Suppose the range of $\lambda - T$ is not closed. Then there exists $x_n \in X$ such that $(\lambda - T)x_n \to y \notin \mathcal{R}(\lambda - T)$. Note $y \neq 0$. Hence, $x_n \notin \ker(\lambda - T)$ for large n so we may assume $x_n \notin \ker(\lambda - T)$ for all n. Since $\ker(\lambda - T)$ is closed, $d_n = dist(x_n, \ker(\lambda - T)) > 0$. Choose $z_n \in \ker(\lambda - T)$ such that $\|x_n - z_n\| < 2d_n$. We claim $\|x_n - z_n\| \to \infty$. If $\|x_n - z_n\| \nrightarrow \infty$, $\{x_n - z_n\}$ contains a bounded subsequence so $\{T(x_n - z_n)\}$ contains a convergent subsequence. But, $x_n - z_n = 1/\lambda((\lambda - T)(x_n - z_n) + T(x_n - z_n))$ and

$$(*) \quad (\lambda - T)(x_n - z_n) = (\lambda - T)x_n \to y$$

so $\{x_n - z_n\}$ must contain a convergent subsequence converging to some $x \in X$. Now $(\lambda - T)x = y$ from $(*)$ and this contradicts the fact that $y \notin \mathcal{R}(\lambda - T)$. Hence, $\|x_n - z_n\| \to \infty$.

Let $w_n = (x_n - z_n)/\|x_n - z_n\|$ so $\|w_n\| = 1$ and

$$(**) \quad (\lambda - T)w_n = \frac{1}{\|x_n - z_n\|}(\lambda - T)x_n \to 0.$$

Now $w_n = \frac{1}{\lambda}((\lambda - T)w_n + Tw_n)$ so from $(**)$ and the fact that T is compact, $\{w_n\}$ has a convergent subsequence with limit, say, w. Since $(\lambda - T)w_n \to 0$, $(\lambda - T)w = 0$. Finally let $y_n = z_n + \|x_n - z_n\| w$. Since both $z_n, w \in \ker(\lambda - T)$, $y_n \in \ker(\lambda - T)$ which implies that $\|x_n - y_n\| \geq d_n$. But,

$$x_n - y_n = x_n - z_n - \|x_n - z_n\| w$$
$$= \|x_n - z_n\| w_n - \|x_n - z_n\| w = \|x_n - z_n\| (w_n - w)$$

so $d_n \leq \|x_n - y_n\| < 2d_n \|w_n - w\|$ which implies $1 < 2\|w_n - w\|$. However, $\{w_n\}$ has a subsequence converging to w. This gives the desired contradiction.

Corollary 15.3. *For every n, $\mathcal{R}((\lambda - T)^n)$ is closed.*

Proof: $(\lambda - T)^n - \lambda^n \quad n\lambda^{n-1}T + \ldots + (-1)^n T^n = \lambda^n - TA$, where A is a continuous linear operator. Then TA is compact and Theorem 2 applies.

Theorem 15.4. $\ker(\lambda - T)$ *is finite dimensional.*

Proof: It suffices to show $S = \{x \in \ker(\lambda - T) : \|x\| \leq 1\}$ is compact (Corollary 4.9). Suppose $x_n \in \ker(\lambda - T), \|x_n\| \leq 1$. Then $x_n = \frac{1}{\lambda}Tx_n$ and since T is compact, $\{x_n\}$ must have a convergent subsequence and S is compact.

Corollary 15.5. $\ker(\lambda - T)^n$ *is finite dimensional for every* $n \geq 1$.

The proof is as in Corollary 3.

Proposition 15.6. *For* $x \in X$ *set* $d(x) = dist(x, \ker(\lambda - T))$. *Then there exists* $M > 0$ *such that* $d(x) \leq M \|(\lambda - T)x\|$ *for* $x \in X$.

Proof: Suppose no such M exists. Then there exists a sequence $\{x_n\}$ such that $d(x_n)/ \|(\lambda - T)x_n\| \to \infty$ with $x_n \notin \ker(\lambda - T)$. Now $\ker(\lambda - T)$ is closed so there exists $u_n \in \ker(\lambda - T)$ such that $d(x_n) = \|x_n - u_n\|$. Set $y_n = (x_n - u_n)/d(x_n)$ so that $\|y_n\| = 1$ and $(\lambda - T)y_n = (\lambda - T)(x_n)/d(x_n) \to 0$. Now $y_n = \frac{1}{\lambda}[(\lambda - T)y_n + Ty_n]$ and since T is compact, $\{y_n\}$ has a subsequence converging to some $y \in X$. Since $(\lambda - T)y_n \to 0$, $(\lambda - T)y = 0$. Then $u_n + d(x_n)y \in \ker(\lambda - T)$ so $\|y_n - y\| = \|x_n - [u_n + d(x_n)y]\| /d(x_n) \geq 1$ contradicting the fact that $\{y_n\}$ has a convergent subsequence.

Proposition 15.7. *Let* M *be as in Proposition 6. If* $y \in \mathcal{R}(\lambda - T)$, *there exists* $x \in X$ *such that* $(\lambda - T)x = y$ *and* $\|x\| \leq M \|y\|$.

Proof: There exists $x_0 \in X$ such that $(\lambda - T)x_0 = y$. Since $\ker(\lambda - T)$ is closed, there exists $w \in \ker(\lambda - T)$ such that $d(x_0) = \|x_0 - w\|$. Set $x = x_0 - w$. Then $(\lambda - T)x = (\lambda - T)x_0 = y$ and by Proposition 6,

$$\|x\| = \|x_0 - w\| = d(x_0) \leq M \|(\lambda - T)x_0\| = M \|y\|.$$

Proposition 15.8. $\mathcal{R}(\lambda - T) = \ker(\lambda - T')_0$ *and* $\mathcal{R}(\lambda - T') = \ker(\lambda - T)^0$.

Proof: $\mathcal{R}(\lambda - T)$ is closed by Theorem 2 so by Proposition 13.14.(2), $\mathcal{R}(\lambda - T) = \ker((\lambda - T)')_0 = \ker(\lambda - T')_0$.

By Theorem 2 and Proposition 13.15.(2), $\mathcal{R}(\lambda - T') \subset \ker(\lambda - T)^0$. Let $y' \in \ker(\lambda - T)^0$. For $y \in \mathcal{R}(\lambda - T)$ define $f(y) = y'(x)$, where $x \in (\lambda - T)^{-1}y$. Since $y' \in \ker(\lambda - T)^0$, f is well defined. For fixed y

choose $x \in (\lambda - T)^{-1}y$ such that $\|x\| \leq M\|y\|$ as in Proposition 7. Then $|f(y)| \leq \|y'\| M \|y\|$ so f is a continuous linear functional on $\mathcal{R}(\lambda - T)$. By the Hahn-Banach Theorem (7.1), f has a continuous linear extension x' on X. Thus, $x'((\lambda - T)x) = y'(x)$ for all $x \in X$ so $y' = x'(\lambda - T) = (\lambda - T')x'$. Hence, $y' \in \mathcal{R}(\lambda - T')$ and $\mathcal{R}(\lambda - T') = \ker(\lambda - T)^0$.

Theorem 15.9. *If $(\lambda - T)$ is one-one, then $(\lambda - T)$ is onto.*

Proof: $\ker(\lambda - T) = \{0\}$ by hypothesis. By Proposition 8, $\mathcal{R}(\lambda - T') = \ker(\lambda - T)^0 = X'$. Thus, $(\lambda - T')$ is onto and by Theorem 1, $(\lambda - T')$ is one-one, i.e., $\ker(\lambda - T') = \{0\}$. By Proposition 8, $\mathcal{R}(\lambda - T) = \ker(\lambda - T')_0 = X$.

Lemma 15.10. *Let Y be a normed space and $f, g_1, ..., g_n \in Y'$ with $g_1, ..., g_n$ linearly independent. Then $f \in span\{g_1, ..., g_n\}$ iff $\ker f \supset \cap_{k=1}^{n} \ker g_k$.*

Proof: If $f \in span\{g_1, ..., g_n\}$, then clearly $\ker f \supset \cap_{k=1}^{n} \ker g_k$.

Suppose $\ker f \supset \cap_{k=1}^{n} \ker g_k$. Define a linear map $G : Y \to \Bbbk^n$ by $G(x) = (g_1(x), ..., g_n(x))$. G is onto by the linear independence assumption. Define $F : \Bbbk^n \to \Bbbk$ by $F(G(x)) = f(x)$. Now F is well defined since if $G(x) = G(x')$, then $x - x' \in \cap_{k=1}^{n} \ker g_k$ so $f(x) = f(x')$ by hypothesis. F is linear so there exists $(s_1, ..., s_n) \in \Bbbk^n$ such that $F(t_1, ..., t_n) = \sum_{k=1}^{n} t_k s_k$ for every $(t_1, ..., t_n) \in \Bbbk^n$. Hence, $f(x) = F(G(x)) = \sum_{k=1}^{n} g_k(x)s_k$ and $f \in span\{g_1, ..., g_n\}$.

Lemma 15.11. *Let Y be a normed space. (1) If $\{y_1, ..., y_n\}$ are linearly independent in Y, there exist $\{y'_1, ..., y'_n\} \subset Y'$ such that $y'_m(y_k) = \delta_{mk}$ for $m, k = 1, ..., n$. (2) If $\{y'_1, ..., y'_n\} \subset Y'$, there exist $\{y_1, ..., y_n\} \subset Y$ such that $y'_m(y_k) = \delta_{mk}$ for $m, k = 1, ..., n$.*

Proof: (1): For $j \in \{1, ..., n\}$ if $M_j = span\{y_i : i \neq j\}$, then M_j is closed and $y_j \notin M_j$ so by Theorem 7.2 there exists $x'_j \in Y'$ such that $x'_j(y_j) \neq 0$ and $x'_j(M_j) = \{0\}$. Set $y'_j = x'_j/x'_j(y_j)$.

(2): For each $j \in \{1, ..., n\}$ there exists $x_j \in Y$ such that $y'_j(x_j) \neq 0$ and $y'_i(x_j) = 0$ for $i \neq j$. For if this is not the case,

$$span\{y'_j\}_0 \supset span\{y'_1, ..., y'_{j-1}, y'_{j+1}, ..., y'_n\}_0,$$

i.e., $\ker\{y'_j\} \supset \cap_{i \neq j} \ker\{y'_i\}$. By Lemma 10 $y'_j \in span\{y'_1, ..., y'_{j-1}, y'_{j+1}, ..., y'_n\}$ contradicting the linear independence assumption.

Theorem 15.12. $\ker(\lambda - T)$ *and* $\ker(\lambda - T')$ *have the same dimension (finite by Theorem 4).*

Proof: Let $m = \dim(\ker(\lambda - T))$, $n = \dim(\ker(\lambda - T'))$. We may assume $m, n \geq 1$ since $m = 0$ iff $n = 0$ by Theorems 1 and 8. Let $\{x_1, ..., x_m\}, \{u'_1, ..., u'_n\}$ be a basis for $\ker(\lambda - T)$ and $\ker(\lambda - T')$, respectively.

By Lemma 11 there exist $\{u_1, ..., u_n\} \subset X$ such that $u'_j(u_i) = \delta_{ij}$ and there exist $\{x'_1, ..., x'_m\} \subset X'$ such that $x'_j(x_i) = \delta_{ij}$.

Suppose $m < n$. Define $A : X \to X$ by $Ax = Tx + \sum_{i=1}^{m} x'_i(x)u_i$. Then A is compact. Applying Proposition 13.14.(1) to $(\lambda - T)$ gives $\mathcal{R}(\lambda - T)^0 = \ker(\lambda - T')$ so

$$(*) \quad u'_j(\lambda - T)x = 0 \ for \ every \ x \in X, j = 1, ..., n.$$

Now

$$(**) \quad (\lambda - A)x = (\lambda - T)x - \sum_{i=1}^{m} x'_i(x)u_i$$

so if $(\lambda - A)x = 0$, then $(\lambda - T)x = \sum_{i=1}^{m} x'_i(x)u_i$ so

$$u'_j(\lambda - T)x = \sum_{i=1}^{m} x'_i(x)u'_j(u_i) = x'_j(x) = 0$$

for all $j = 1, ..., n$ by (*). Thus, if $(\lambda - A)x = 0$, then $(\lambda - T)x = 0$ since $(\lambda - T)x = (\lambda - A)x + \sum_{i=1}^{m} x'_i(x)u_i = 0$ and $x \in \ker(\lambda - T)$ so $x = \sum_{i=1}^{m} t_i x_i$ and $x'_j(x) = \sum_{i=1}^{m} t_i x'_j(x_i) = t_j = 0$ which implies $x = 0$. Thus, $(\lambda - A)$ is one-one and, therefore, onto by Theorem 9.

Choose x such that $(\lambda - A)x = u_{m+1}$ (recall $n > m$). Now $u'_{m+1}(u_{m+1}) = 1$, but by (**) and (*)

$$1 = u'_{m+1}(u_{m+1}) = u'_{m+1}((\lambda - A)x) = u'_{m+1}((\lambda - T)x) - \sum_{i=1}^{m} x'_i(x)u'_{m+1}(u_i) = 0$$

an obvious contradiction.

Suppose $m > n$. define $B \in L(X', X')$ by $Bx' = T'x' + \sum_{i=1}^{n} x'(u_i)x'_i$ and proceed as above.

Theorem 15.13. *(Fredholm Alternative) Either the homogeneous equation $\lambda x - Tx = 0$ has only the trivial solution in which case the non-homogeneous equation $\lambda x - Tx = y$ has a unique solution for every $y \in X$ OR the homogeneous equation has a finite number $\nu = \dim \ker(\lambda - T)$ of linearly independent solutions in which case the conjugate homogeneous equation $\lambda x' - T'x' = 0$ has ν linearly independent solutions; the non-homogeneous equation has solutions for those $y \in \ker(\lambda - T')_0$.*

The statement in the Fredholm Alternative above is directly applicable to the Fredholm integral equation

$$g(t) = f(t) - \lambda \int_a^b k(t,s)f(s)ds,$$

when $k \in C([a,b] \times [a,b]), g \in C[a,b]$.

Exercises.

1. Show the conclusion of Theorem 9 is false if $\lambda = 0$. [Hint: Use the Volterra operator.]

2. Show Theorems 1 and 9 are false if the operator is not compact.

3. Show the conclusion in Theorem 4 is false if $\lambda = 0$. [Hint: Use the kernel $k(t,s) = t \sin s$.]

4. Use the Fredholm Alternative to show that the equation

$$f(t) - \int_0^\pi \sin(t+s)f(s)ds = g(t)$$

has a unique solution for every $g \in C[0,\pi]$.

Chapter 16

The Spectrum of an Operator

The spectrum of a linear operator on a finite dimensional space is just the set of eigenvalues of the operator. In infinite dimensional spaces the situation is more complicated; for example, the compact Volterra operator $Tf(t) = \int_0^t f(s)ds$ acting on $C[0,1]$ has no eigenvalues. In this chapter we define the spectrum of a linear operator and study its properties. Examples are then given in the next chapter.

Let $X \neq \{0\}$ be a complex normed space; as in the finite dimensional case it is important to use complex scalars. Let T be a linear operator with domain $\mathcal{D}T \subset X$ and with range in X.

Definition 16.1. If $\lambda \in \mathbb{C}$ is such that $\lambda - T = \lambda I - T$ has a dense range in X and a continuous inverse (on $\mathcal{R}T$) , then λ is said to belong to the resolvent set of T. The resolvent set is denoted by $\rho(T)$. The complement of the resolvent set is called the spectrum of T and is denoted by $\sigma(T)$.

For continuous linear operators acting on Banach spaces, the resolvent has a simpler description.

Theorem 16.2. Let X be a Banach space and $T \in L(X)$. Then $\lambda \in \rho(T)$ iff $(\lambda - T)^{-1} \in L(X)$.

Proof: If $(\lambda - T)^{-1} \in L(X)$, then clearly $\lambda \in \rho(T)$.

Suppose $\lambda \in \rho(T)$. Then $\lambda - T$ has a continuous inverse so $\mathcal{R}(\lambda - T)$ is closed since X is complete (Exercise 13.2). Since $\mathcal{R}(\lambda - T)$ is dense, $\lambda - T$ is onto.

Thus, if X is finite dimensional, the spectrum of a linear operator is just the set of eigenvalues of the operator. We proceed to establish properties of the spectrum.

Lemma 16.3. *Suppose $\mu \in \mathbb{C}$ is such that $\mu - T$ has a bounded inverse and let $M(\mu) = \left\| (\mu - T)^{-1} \right\|$ (the norm as a continuous linear operator on the range of $\mu - T$). If $\lambda \in \mathbb{C}$ is such that $|\lambda - \mu| \, M(\mu) < 1$, then $\lambda - T$ has a bounded inverse and $\overline{\mathcal{R}(\lambda - T)}$ is not a proper subset of $\overline{\mathcal{R}(\mu - T)}$.*

Proof: Let $x \in \mathcal{D}T$. Then $(\lambda - T)x = (\lambda - \mu)x + (\mu - T)x$ implies $\| (\lambda - T)x \| \geq \| (\mu - T)x \| - |\lambda - \mu| \, \|x\|$. Since

$$\|x\| = \left\| (\mu - T)^{-1}(\mu - T)x \right\| \leq M(\mu) \, \| (\mu - T)x \| ,$$

$$
\begin{aligned}
M(\mu) \, \| (\lambda - T)x \| &\geq M(\mu) \, \| (\mu - T)x \| - M(\mu) \, |\lambda - \mu| \, \|x\| \\
&\geq \|x\| \, (1 - M(\mu) \, |\lambda - \mu|).
\end{aligned}
$$

Since $1 - M(\mu) \, |\lambda - \mu| > 0$, Theorem 2.11 implies $\lambda - T$ has a continuous inverse.

Suppose $\overline{\mathcal{R}(\lambda - T)}$ is a proper subset of $\overline{\mathcal{R}(\mu - T)}$. Choose θ such that $|\lambda - \mu| \, M(\mu) < \theta < 1$. By Riesz's Lemma (4.6) there exists $y_\theta \in \overline{\mathcal{R}(\mu - T)}$, $\|y_\theta\| = 1$, such that $\|y - y_\theta\| \geq \theta$ for all $y \in \overline{\mathcal{R}(\lambda - T)}$. Choose $y_n \in \mathcal{R}(\mu - T)$ such that $\|y_n - y_\theta\| \to 0$. Set $x_n = (\mu - T)^{-1} y_n$. Since $(\lambda - T)x_n \in \mathcal{R}(\lambda - T)$, we have

$$\theta \leq \| (\lambda - T)x_n - y_\theta \| \leq \| (\mu - T)x_n - y_\theta \| + \| (\lambda - T)x_n - (\mu - T)x_n \|$$
$$= \|y_n - y_\theta\| + |\lambda - \mu| \, \|x_n\| \leq \|y_n - y_\theta\| + |\lambda - \mu| \, M(\mu) \, \|y_n\| .$$

Letting $n \to \infty$ gives $\theta \leq |\lambda - \mu| \, M(\mu) \, \|y_\theta\| = |\lambda - \mu| \, M(\mu)$ which is a contradiction.

Theorem 16.4. *The resolvent set $\rho(T)$ is open and the spectrum $\sigma(T)$ is closed.*

Proof: If $\mu \in \rho(T)$, then $(\mu - T)^{-1}$ is continuous and $\overline{\mathcal{R}(\mu - T)} = X$. Lemma 3 implies that if λ is close to μ , then $\lambda - T$ has a continuous inverse and $\overline{\mathcal{R}(\lambda - T)}$ is not a proper subset of X so $\mathcal{R}(\lambda - T)$ must be dense in X, i.e., $\lambda \in \rho(T)$.

For $\lambda \in \rho(T)$, we write $R_\lambda(T) = R_\lambda = (\lambda - T)^{-1}$; the operator R_λ is called the *resolvent operator* of T. We have the following important property of the resolvent operator.

Theorem 16.5. *Let X be a Banach space and $T \in L(X)$. For $\lambda, \mu \in \rho(T)$ we have (1) $R_\lambda - R_\mu = (\mu - \lambda) R_\lambda R_\mu$ and (2) $R_\lambda R_\mu = R_\mu R_\lambda$.*

Proof: $(\mu - T)(\lambda - T)(R_\lambda - R_\mu) = \mu - T - (\lambda - T) = (\mu - \lambda)I$. Multiplying this equation by $R_\lambda R_\mu$ gives (1). By symmetry $R_\mu - R_\lambda = (\lambda - \mu) R_\mu R_\lambda$ and adding this to (1) gives (2).

The equation in (1) is called the *resolvent equation*.

Theorem 16.6. *Let X be a Banach space and $T \in L(X)$. If $|\lambda| > \lim \sqrt[n]{\|T^n\|}$ (recall this limit exists), then $(\lambda - T)^{-1} \in L(X)$ and $R_\lambda = (\lambda - T)^{-1} = \sum_{n=1}^{\infty}(1/\lambda^n)T^{n-1}$ (where the series is norm convergent in $L(X)$) and $\|R_\lambda\| \leq 1/(|\lambda| - \|T\|)$.*

Proof: Since $\lambda - T = \lambda(I - T/\lambda)$ and $\lim \sqrt[n]{\|(T/\lambda)^n\|} = (1/|\lambda|) \lim \sqrt[n]{\|T^n\|} < 1$ it follows from Corollary 2.16 that the Neumann series for $(I - T/\lambda)^{-1} = \sum_{n=1}^{\infty} T^n/\lambda^n$ converges in $L(X)$ so $(\lambda - T)^{-1} = \sum_{n=1}^{\infty}(1/\lambda^n)T^{n-1}$. For the last inequality,

$$\|(\lambda - T)^{-1}\| = \left\|\sum_{n=1}^{\infty}(1/\lambda^n)T^{n-1}\right\| \leq (1/|\lambda|)\sum_{n=0}^{\infty}\|T/\lambda\|^n = 1/(|\lambda| - \|T\|).$$

Corollary 16.7. *Let X be a Banach space and $T \in L(X)$. The spectrum of T, $\sigma(T)$, is compact with $\sigma(T) \subset \{\lambda : |\lambda| \leq \lim \sqrt[n]{\|T^n\|}\}$.*

Proof: $\sigma(T)$ is closed by Theorem 4 and is bounded by $\lim \sqrt[n]{\|T^n\|}$ from Theorem 6.

We now show that the spectrum of a continuous linear operator on a complex Banach space is non-empty. The proof uses Liouville's Theorem from complex analysis ([Ah],[CB]).

Theorem 16.8. *Let X be a Banach space and $T \in L(X)$. Then $\sigma(T) \neq \emptyset$.*

Proof: Let $z \in L(X)'$. Define $f = f_z : \rho(T) \to \mathbb{C}$ by $f(\lambda) = z(R_\lambda) = z((\lambda - T)^{-1})$. By the resolvent equation

$$(f(\lambda) - f(\mu))/(\lambda - \mu) = z((R_\lambda - R_\mu)/(\lambda - \mu)) = z(-R_\mu R_\lambda).$$

By Corollary 2.18 the map $\lambda \to R_\lambda$ from $\rho(T)$ to $L(X)$ is continuous so letting $\lambda \to \mu$ gives $\lim_{\lambda \to \mu}(f(\lambda) - f(\mu))/(\lambda - \mu) = -z(R_\mu^2)$. Thus, f is analytic on $\rho(T)$ with $f'(\mu) = -z(R_\mu^2)$.

Moreover, f is bounded for large λ since $|f(\lambda)| \leq \|z\| \|R_\lambda\| \leq \|z\|/(|\lambda| - \|T\|)$ for $|\lambda| > \|T\|$ by Theorem 6. Hence, if $\sigma(T) = \emptyset$, then f would be a bounded entire function and by Liouville's Theorem, f would be constant. Since $f(\lambda) \to 0$ as $|\lambda| \to \infty$, f would have to be 0. Since $z \in L(X)'$ is arbitrary, this would imply that $R_\lambda = 0$ which is not possible since R_λ is an inverse.

Corollary 16.9. *Let X be a Banach space and $T \in L(X)$. For $z \in L(X)'$, the function $f(\lambda) = z(R_\lambda) = z((\lambda - T)^{-1})$ is analytic on $\rho(T)$ with $f'(\lambda) = -z(R_\lambda^2)$.*

From Theorem 8 and Corollary 7, we have that $\sigma(T)$ is a non-empty compact subset of $\mathbb{C}$. The operator in Example 17.6 shows that any compact subset of $\mathbb{C}$ is the spectrum of a continuous linear operator on a Hilbert space.

Definition 16.10. For $T \in L(X)$ the spectral radius of T is defined to be $r(T) = \sup\{|\lambda| : \lambda \in \sigma(T)\}$.

Theorem 16.11. *Let X be a Banach space and $T \in L(X)$. Then $r(T) \leq \|T^n\|^{1/n}$ for every $n \geq 1$ and $r(T) = \lim \sqrt[n]{\|T^n\|}$.*

Proof: From Corollary 7, we have $r(T) \leq \lim \sqrt[n]{\|T^n\|}$. The series $\sum_{n=0}^{\infty}(1/\lambda^{n+1})T^n$ converges in $L(X)$ to R_λ for $|\lambda| > \|T\|$. From Corollary 9, for every $z \in L(X)'$, the series $\sum_{n=0}^{\infty}(1/\lambda^{n+1})z(T^n)$ converges to the analytic function $-z(R_\lambda^2)$ for $|\lambda| > r(T)$. Therefore, for $|\lambda| > r(T)$, $\sup\{|z(T^n)/\lambda^{n+1}| : n\} < \infty$ so by the Uniform Boundedness Principle, $\sup\{\|T^n/\lambda^{n+1}\| : n\} = M < \infty$. Hence, $\|T^n\| \leq M|\lambda^{n+1}|$ and $\lim \sqrt[n]{\|T^n\|} \leq \lim(M|\lambda|^{n+1})^{1/n} = |\lambda|$ for $|\lambda| > r(T)$ so $\lim \sqrt[n]{\|T^n\|} \leq r(T)$.

Theorem 16.12. *(Spectral Mapping Theorem: Junior Grade) Let X be a Banach space and $T \in L(X)$. If p is a complex polynomial, then $\sigma(p(T)) = p(\sigma(T))$.*

Proof: Fix $\mu \in \mathbb{C}$ and let $p(z) - \mu = c \prod_{j=1}^{n} (z - b_j)$ so

$$(*) \quad p(T) - \mu = c \prod_{j=1}^{n} (T - b_j).$$

If $\mu \in \sigma(T)$, then by (*) some $b_j \in \sigma(T)$ since otherwise $p(T) - \mu$ would be invertible. But, $p(b_j) = \mu$ so $\mu \in p(\sigma(T))$ and $\sigma(p(T)) \subset p(\sigma(T))$.

Suppose some $b_j \in \sigma(T)$, say, b_1. If $T - b_1$ has a bounded inverse, then the range of $T - b_1$ is not X and (*) implies the range of $p(T) - \mu$ is not X. Then $\mu \in \sigma(p(T))$. If $T - b_1$ does not have a bounded inverse, then (*) implies that $p(T) - \mu$ has no inverse so in this case $\mu \in \sigma(p(T))$. Hence, $\sigma(p(T)) \supset p(\sigma(T))$.

Exercises.

1. Suppose X is a Banach space with $X = M_1 \oplus M_2$ and with M_i a closed subspace. If $T \in L(X)$ is such that $TM_i \subset M_i$, set $T_i = T |_{M_i}$. Show $\sigma(T) = \sigma(T_1) \cup \sigma(T_2)$.

2. Let X be a Banach space and $T \in L(X)$. Show $\sigma(T) = \sigma(T')$.

Chapter 17

Subdivisions of the Spectrum

In this chapter we will define the subdivisions of the spectrum and give examples illustrating the divisions.

Let $X \neq \{0\}$ be a complex normed space and $T : \mathcal{D}(T) \subset X \to X$ be a linear operator.

Definition 17.1. A point $\lambda \in \sigma(T)$ is an eigenvalue if there exists $x \neq 0$ such that $Tx = \lambda x$ and such a vector x is called an eigenvector associated with the eigenvalue λ. The collection of all eigenvalues of T is called the point spectrum of T and is denoted by $P\sigma(T)$.

Definition 17.2. All $\lambda \in \sigma(T)$ such that $\lambda - T$ is one-one and $\mathcal{R}(\lambda - T)$ is dense in X is called the continuous spectrum of T and is denoted by $C\sigma(T)$.

Definition 17.3. All $\lambda \in \sigma(T)$ such that $\lambda - T$ is one-one but $\mathcal{R}(\lambda - T)$ is not dense in X is called the residual spectrum of T and is denoted by $R\sigma(T)$.

Thus, the spectrum of T is the disjoint union of $P\sigma(T)$, $C\sigma(T)$, and $R\sigma(T)$.

We now give several examples of spectrums and their divisions.

Example 17.4. $X = \mathbb{C}^n$. If $T : X \to X$ is linear, then $\sigma(T) = P\sigma(T)$.

Example 17.5. Suppose $X = C[0,1]$ (complex scalars) and $T : X \to X$ is defined by $Tf(t) = tf(t)$. Then $\|T\| = 1$ so $\sigma(T) \subset \{\lambda : |\lambda| \leq 1\}$. Suppose $\lambda = a + bi$, $|\lambda| \leq 1$, with $b \neq 0$. Then
$$|(\lambda - T)f(t)|^2 = |(a-t)f(t) + ibf(t)|^2 = (a-t)^2|f(t)|^2 + b^2|f(t)|^2 \geq b^2|f(t)|^2$$
which implies that $\|(\lambda - T)f\| \geq |b| \|f\|$ so $\lambda \in \rho(T)$ since $\lambda - T$ has a continuous inverse (2.11) and is clearly onto X. Thus, $\sigma(T) \subset [-1,1]$.

Suppose $-1 \le \lambda < 0$. Then $|(\lambda - T)f(t)| = |\lambda - t|\,|f(t)| \ge |\lambda|\,|f(t)|$ which implies $\|(\lambda - T)f\| \ge |\lambda|\,\|f\|$ so $\lambda \in \rho(T)$ since $\lambda - T$ has a continuous inverse and is clearly onto X. Thus, $\sigma(T) \subset [0,1]$. Let $\lambda \in [0,1]$. Let $\epsilon > 0$ be such that either $[\lambda, \lambda + \epsilon]$ or $[\lambda - \epsilon, \lambda]$ is contained in $[0,1]$ (suppose the former). Construct $f_\epsilon \in X$ to be equal to 1 on $[\lambda + \epsilon/3, \lambda + 2\epsilon/3]$, 0 outside $[\lambda, \lambda + \epsilon]$ and linear on $[\lambda, \lambda + \epsilon/3]$ and $[\lambda + 2\epsilon/3, \lambda]$. Then $\|f_\epsilon\| = 1$ and $\epsilon \ge \|(\lambda - T)f_\epsilon\|$ so $\|(\lambda - T)f_\epsilon\| \to 0$ as $\epsilon \to 0$. Therefore, $(\lambda - T)$ doesn't have a continuous inverse (Theorem 2.11) and $\lambda \in \sigma(T)$. Thus, $\sigma(T) = [0,1]$. Note that $P\sigma(T) = \emptyset$ since $Tf(t) = tf(t) = \lambda f(t)$ for all t implies $f(t) = 0$ for $\lambda \ne t$ and, therefore, $f = 0$. Note also, $\mathcal{R}(\lambda - T)$ is not dense in X since $(\lambda - T)f$ vanishes at $\lambda = t$ for all $f \in X$. Thus, $\sigma(T) = R\sigma(T) = [0,1]$.

A point λ is called an *approximate eigenvalue* of T if to each $\epsilon > 0$ there corresponds an $x = x_\epsilon \in \mathcal{D}(T)$ such that $\|x\| = 1$ and $\|\lambda - T)x\| < \epsilon$. By Theorem 2.11 a point λ is an approximate eigenvalue iff $(\lambda - T)$ does not have a continuous inverse so an approximate eigenvalue belongs to the spectrum of T. The set of all approximate eigenvalues of T is called the *approximate spectrum* of T. Note that in Example 5 every point of the spectrum is an approximate eigenvalue.

Example 17.6. Let $K \ne \emptyset$ be a compact subset of $\mathbb{C}$ and let $D = \{d_j : j \in \mathbb{N}\}$ be a dense subset of K. Define $T : X = l^2 \to l^2$ by $T(\{t_j\}) = \{d_j t_j\}$. If $\lambda \in \mathbb{C} \setminus K$, there exists $d > 0$ such that $|\lambda - d_j| > d$ for all j. Then $\|(\lambda - T)\{t_j\}\|^2 = \sum_{j=1}^{\infty} |(\lambda - d_j)t_j|^2 \ge d^2 \sum_{j=1}^{\infty} |t_j|^2$ so $\lambda - T$ has a bounded inverse and since $\lambda - T$ is onto, $\lambda \in \rho(T)$. Hence, $\sigma(T) \subset K$. Since $Te^j = d_j e^j$, each d_j is an eigenvalue with e^j as an associated eigenvector. Hence, $K = \overline{D} = \sigma(T)$. If $\lambda \in K \setminus D$, we have $(\lambda - T)\{t_j\} = 0$ implies $(\lambda - d_j)t_j = 0$ for all j so $\{t_j\} = 0$ and $\lambda - T$ is one-one and $P\sigma(T) = D$. Note $\lambda - T$ has dense range since the range of $\lambda - T$ contains $\{e^j\}$ $[(\lambda - T)(e^j/(\lambda - d_j)) = e^j]$. Therefore, $C\sigma(T) = K \setminus D$ and $R\sigma(T) = \emptyset$.

This example shows that any compact subset of the complex plane is the spectrum of a continuous linear operator on the Hilbert space l^2.

We now consider 2 examples which show that the extremes for the spectrum can occur.

Example 17.7. Let $X = C[0,1]$, where we are using complex scalars. Let $\mathcal{D}(T) = \{f \in X : f' \in X\}$ and define $D : \mathcal{D}(T) \to X$ by $Df = f'$. Then $\sigma(T) = P\sigma(T) = \mathbb{C}$ since any $\lambda \in \mathbb{C}$ is an eigenvalue with associated eigenfunction $f(t) = e^{\lambda t}$.

Example 17.8. Let $X = C[0,1]$, where we are using complex scalars. Let $\mathcal{D}(T) = \{f \in X : f' \in X, f(0) = 0\}$ and define $D : \mathcal{D}(T) \to X$ by $Df = f'$. For each $\lambda \in \mathbb{C}$ the operator R_λ defined by $R_\lambda f(t) = e^{-\lambda t} \int_0^t f(s) e^{\lambda s} ds$ belongs to $L(X)$ and $R_\lambda(\lambda - T) = (\lambda - T)R_\lambda = I$ so R_λ is a resolvent operator. Hence, $\rho(T) = \mathbb{C}$ and $\sigma(T) = \emptyset$.

This example shows that the completeness assumption in Theorem 16.8 is important.

Example 17.9. Let L be the left shift operator on l^p for $1 \le p < \infty$. $[L(\{t_j\}) = (t_2, t_3, ...)]$. If $|\lambda| < 1$, then λ is an eigenvalue with associated eigenvector $(1, \lambda, \lambda^2, ...)$. Hence, $P\sigma(L) \supset \{\lambda : |\lambda| < 1\}$ and $\sigma(L) = \{\lambda : |\lambda| \le 1\}$ since $\|L\| = 1$. If $|\lambda| = 1$, then $\lambda - L$ is one-one since $(1, \lambda, \lambda^2, ...) \notin l^p$. The range of $\lambda - L$ is dense in l^p since $(\lambda - L)(e^j - (\lambda - 1)e^{j+1}) = e^j$. Hence, $C\sigma(L) = \{\lambda : |\lambda| = 1\}$. ·

The reader is asked to repeat Example 9 with $p = \infty$ in Exercise 1 and asked to consider the right shift operator in Exercise 2.

Exercises.

1. Repeat Example 9 for the case when $p = \infty$.
2. Let R be the right shift operator $R(\{t_j\}) = (0, t_1, t_2, ...)$. Find and classify the spectrum of R on l^p for $1 \le p \le \infty$.
3. In Exercise 16.1, show $P\sigma(T) = P\sigma(T_1) \cup P\sigma(T_2)$.
4. Let $C_\infty(\mathbb{R})$ be the space of complex valued functions on $\mathbb{R}$ which vanish at ∞ equipped with the sup-norm. Let $s \in \mathbb{R}$ and define the shift operator $S : C_\infty(\mathbb{R}) \to C_\infty(\mathbb{R})$ by $Sf(t) = f(t+s)$. Show $\sigma(S) = \{z : |z| = 1\}$.
5. Let $BC(\mathbb{R})$ be the space of all bounded, continuous, complex valued functions on $\mathbb{R}$ with the sup-norm. Fix $s \in \mathbb{R}$ and define $Sf(t) = f(t+s)$ for $t \in \mathbb{R}, f \in BC(\mathbb{R})$. Show S is an isometry onto $BC(\mathbb{R})$ and find $\sigma(S)$.

Chapter 18

The Spectrum of a Compact Operator

In this chapter we study the spectrum of a compact operator. The spectrum of these operators is much simpler to describe than that of an arbitrary continuous linear operator. In a later chapter we will give applications to Sturm-Liouville differential operators.

Let $X \neq \{0\}$ be a complex normed space and $T \in K(X, X)$.

Theorem 18.1. *Let $0 \neq \lambda \in \mathbb{C}$. If $M > 0$ is the constant in Proposition 15.7, then to each $y \in \mathcal{R}(\lambda - T)$ there exists $x \in X$ such that $(\lambda - T)x = y$ and $\|x\| \leq M \|y\|$. Thus, if $(\lambda - T)^{-1}$ exists, then $(\lambda - T)^{-1}$ is continuous. The point λ is either in $P\sigma(T)$ or in $\rho(T)$.*

Proof: If $\lambda \in \sigma(T) \setminus P\sigma(T)$, then $(\lambda - T)$ is one-one and, therefore, onto by the Fredholm Alternative (Theorem 15.13). Thus, $(\lambda - T)^{-1} \in L(X)$ and this is a contradiction.

Theorem 18.2. *Let $0 \neq \lambda \in \mathbb{C}$. Then $\ker(\lambda - T)$ is finite dimensional.*

Proof: It suffices to show that $S = \{x \in \ker(\lambda - T) : \|x\| \leq 1\}$ is compact (Corollary 4.9). Let $\{x_j\} \subset S$. Then $x_j = (1/\lambda)Tx_j$ and since T is compact, $\{Tx_j\}$ has a convergent subsequence so the same is true for $\{x_j\}$. Hence, S is compact.

Lemma 18.3. *Let $\{\lambda_j\}$ be a distinct sequence of eigenvalues of T with x_j an eigenvector associated with λ_j. Then $\lambda_j \to 0$.*

Proof: Suppose $\{\lambda_j\}$ does not converge to 0. Then there exists $\epsilon > 0$ such that $|\lambda_j| \geq \epsilon$ for infinitely many j; for convenience assume this holds for all j. Let $H_k = span\{x_1, ..., x_k\}$. So H_k is a closed subspace of X. We claim that $H_k \subsetneq H_{k+1}$. For this it suffices to show that $\{x_1, ..., x_k\}$ is linearly independent for each k. Suppose $\{x_1, ..., x_{n-1}\}$ is linearly independent but $x_n = \sum_{k=1}^{n-1} t_k x_k$. Then

$$0 = (\lambda_n - T)x_n = \sum_{k=1}^{n-1} t_k(\lambda_n - T)x_k = \sum_{k=1}^{n-1} t_k(\lambda_n - \lambda_k)x_k$$

which implies that $t_k = 0$ for $k = 1, ..., n-1$ since $(\lambda_n - \lambda_k) \neq 0$ for $k = 1, ..., n-1$. This implies $x_n = 0$ which is impossible and the claim is established.

By Riesz's Lemma (4.6), for every n there exists $y_n \in H_n$ such that $\|y_n\| = 1$ and $\|y_n - x\| \geq 1/2$ for every $x \in H_{n-1}$. Each y_n has the form $y_n = \sum_{k=1}^{n} t_k x_k$ so $(\lambda_n - T)y_n \in H_{n-1}$. Thus, if $n > m$, then $z_{nm} = (y_n - \frac{1}{\lambda_n}Ty_n) + \frac{1}{\lambda_m}Ty_m \in H_{n-1}$ and $\left\|T(\frac{1}{\lambda_n}y_n) - T(\frac{1}{\lambda_m}y_m)\right\| = \|y_n - z_{nm}\| \geq 1/2$. Hence, no subsequence of $\{T(\frac{1}{\lambda_n}y_n)\}$ converges while $\left\|\frac{1}{\lambda_n}y_n\right\| \leq 1/\epsilon$. This contradicts the compactness of T.

Theorem 18.4. *The spectrum of T is at most countable and has no point of accumulation except possibly 0.*

Proof: Since $\sigma(T)$ is compact, it suffice to show that every $\lambda \neq 0$ in the spectrum of T is an isolated point. If such a λ is not an isolated point of the spectrum, then there exist $0 \neq \lambda_k \in \sigma(T)$ such that $\lambda_k \to \lambda$ with $\{\lambda_k\}$ distinct. By Theorem 1 each λ_k is an eigenvalue and Lemma 3 implies that $\lambda_k \to 0$; a contradiction.

The operator in Example 17.6 can be used to show that the range of any null sequence $\{\lambda_j\} \subset \mathbb{C}$ plus $\{0\}$ can be the spectrum of a compact operator on the Hilbert space l^2.

Example 18.5. Let $\lambda_j \to 0$ in $\mathbb{C}$. Consider the operator $T : l^2 \to l^2$ defined by $T(\{t_j\}) = \{\lambda_j t_j\}$ (Example 17.6). Then $Te^j = \lambda_j e^j$ so $P\sigma(T) = \{\lambda_j : j \in \mathbb{N}\}$ and $\sigma(T) = \{\lambda_j : j \in \mathbb{N}\} \cup \{0\}$. Note that T is compact since if $T_n\{t_j\} = \sum_{j=1}^{n} \lambda_j t_j e^j$, then $\|(T_n - T)\{t_j\}\|^2 \leq \sup\{|\lambda_j|^2 : j \geq n+1\}\sum_{j=1}^{\infty} |t_j|^2$ (Corollary 14.6).

Theorem 18.6. *If X is infinite dimensional , then $0 \in \sigma(T)$.*

Proof: If $0 \in \rho(T)$, then $T^{-1} \in L(X)$ and $I = T^{-1}T$ is compact. Hence, X is finite dimensional (Corollary 4.9).

The Volterra operator gives an example of a compact operator with only $\{0\}$ as its spectrum (Exercise 2).

We give examples of compact operators and their spectrums.

Example 18.7. Let h be a complex valued, periodic function with period 2π and let H be the compact integral operator on $C[-\pi, \pi]$ induced by h and defined by $Hf(t) = \int_{-\pi}^{\pi} h(t - s)f(s)ds$. Let $f_k(t) = \frac{1}{\sqrt{2\pi}}e^{ikt}$ for $k = 0, \pm 1, \pm 2, \ldots$. Then $\{f_k : k = 0, \pm 1, \pm 2, \ldots\}$ is an orthonormal subset in $C[-\pi, \pi]$. For each k we have

$$Hf_k(t) = \int_{-\pi}^{\pi} h(t - s)f_k(s)ds = \int_{t-\pi}^{t+\pi} h(s)f_k(t - s)ds$$
$$= \frac{1}{\sqrt{2\pi}}e^{ikt}\int_{-\pi}^{\pi} h(s)e^{-iks}ds = \lambda_k f_k(t),$$

where $\lambda_k = \int_{-\pi}^{\pi} h(s)e^{-iks}ds$. Therefore, $\{\lambda_k\}$ are eigenvalues of H with associated eigenvectors $\{f_k\}$. Notice that λ_k is the kth Fourier coefficient of the function h with respect to the orthonormal subset $\{f_k\}$. Therefore, the $\{\lambda_k\}$ converge to 0 by the Riemann-Lebesgue Lemma ([HS]16.35,[Sw1]Exercise 6.6.6).

Example 18.8. Let $k(s, t) = \cos(s - t)$ and K be the compact integral operator on the interval $[0, 2\pi]$ with kernel k. We describe the eigenvalues of K. The equation for eigenvalues is

$$(*) \int_0^{2\pi} \cos(s - t)u(s)ds = \lambda u(t) = \cos t \int_0^{2\pi} \cos su(s)ds + \sin t \int_0^{2\pi} \sin su(s)ds.$$

If $\lambda \neq 0$, this equation implies that an eigenfunction associated with λ must be a linear combination of $\cos t$ and $\sin t$. If $u(t) = a\cos t + b\sin t$, substituting into $(*)$ gives $\pi a = \lambda a$, $\pi b = \lambda b$ so $\lambda = \pi$ and any eigenfunction associated with $\lambda = \pi$ must be a linear combination of $\cos t, \sin t$. Thus, the eigenspace of $\lambda = \pi$ is 2 dimensional. Equation $(*)$ also implies that 0 is an eigenvalue of K and the eigenspace of $\lambda = 0$ is the orthogonal complement of $\{\cos t, \sin t\}$ and has infinite multiplicity.

If $M \subset X$ is a proper subspace and $T \in L(X)$, then M is *invariant* under T if $TM \subset M$. If $\lambda \in P\sigma(T)$, then T has an invariant subspace. It is known that any compact operator has an invariant subspace, but it is an old problem in operator theory as to whether any continuous linear operator on a Hilbert space has an invariant subspace. There exist examples of continuous linear operators on Banach spaces with no invariant subspaces. See [AA] for a discussion.

Exercises.

1. Let $a_j \to 0$. Define $T : l^2 \to l^2$ by $T\{t_j\} = (0, a_1 t_2, a_2 t_3, ...)$. Show that T is compact and compute $P\sigma(T), \|T\|$ and $r(T)$.

2. If T is the Volterra operator, show $\sigma(T) = \{0\}$. Show the same holds for the general Volterra operator of Exercise 14.1.

3. Give an example of an invariant subspace for the Volterra operator. (Hint: Consider $C_a = \{f : f = 0 \text{ on } [0, a]\}$.)

Chapter 19

Symmetric Linear Operators

Let X be a complex inner product space and $T : \mathcal{D}(T) \subset X \to X$ be a linear operator.

Definition 19.1. The operator T is symmetric if $Tx \cdot y = x \cdot Ty$ for every $x, y \in \mathcal{D}(T)$.

Example 19.2. If X is a Hilbert space and $T \in L(X)$ is such that $T = T^*$ (i.e., T is self adjoint), then T is symmetric. This is immediate since $Tx \cdot x = x \cdot T^*x = x \cdot Tx$ by the definition of the adjoint operator T^*.

It is a remarkable result of Hellinger and Toeplitz that any symmetric operator on a Hilbert space is continuous and, therefore, self adjoint. Such results are referred to as an automatic continuity result since an algebraic condition such as symmetry implies a continuity result.

Theorem 19.3. *(Hellinger-Toeplitz) Let X be a Hilbert space and $T : X \to X$ be symmetric. Then T is continuous.*

Proof: The set $\{Tx : \|x\| \leq 1\}$ is weakly bounded in X since for every $y \in X$, $|Tx \cdot y| = |x \cdot Ty| \leq \|Ty\|$. Therefore, by the Uniform Boundedness Principle, $\sup\{\|Tx\| : \|x\| \leq 1\} = \|T\| < \infty$.

The reader is asked to supply another proof using the Closed Graph Theorem in Exercise 1. Exercise 2 shows that the completeness assumption in the Hellinger-Toeplitz Theorem is important.

Example 19.4. Let $k \in C([a, b] \times [a, b])$ and let K be the integral operator on $C[a, b]$ induced by the kernel k. Give $C[a, b]$ the inner product $f \cdot g = \int_a^b f\overline{g}$. Then K is symmetric iff $k(s, t) = \overline{k(t, s)}$ for $s, t \in [a, b]$.

Example 19.5. Let X be as in Example 4. Define $Lf = -(pf')' + qf$, where p and q are real valued functions with $p \in C^1[a,b]$ and $q \in C[a,b]$. We consider the boundary conditions $B_1 f = a_1 f(a) + a_2 f'(a) = 0$ and $B_2 f = b_1 f(b) + b_2 f'(b) = 0$, where $a_1^2 + a_2^2 > 0, b_1^2 + b_2^2 > 0$ and the a_i, b_i are real. The domain of L is defined to be $\mathcal{D}(L) = \{f \in C^2[a,b] : B_1 f = 0, B_2 f = 0\}$. The linear operator L is called a Sturm-Liouville operator. We show that L is symmetric. Let $f, g \in \mathcal{D}(L)$. Then

$Lf \cdot g - f \cdot Lg = \int_a^b \{[-(pf')' + qf]\overline{g} - qf\overline{g} + (p\overline{g'})'f\}$. Integrating by parts gives

$$(*) \quad Lf \cdot g - f \cdot Lg = p(b)[f(b)\overline{g'}(b) - f'(b)\overline{g}(b)] - p(a)[f(a)\overline{g'}(a) - f'(a)\overline{g}(a)].$$

Since a_i and b_i are real, $\overline{f}, \overline{g}$ belong to $\mathcal{D}(L)$. Applying B_1 to f and $\overline{g}$ gives $a_1 f(a) + a_2 f'(a) = 0, a_1 \overline{g}(a) + a_2 \overline{g'}(a) = 0$, and since not both a_1 and a_2 are 0, $\det \begin{bmatrix} f(a) & f'(a) \\ \overline{g}(a) & \overline{g'}(a) \end{bmatrix} = 0$. Thus, the second term on the right hand side of $(*)$ is 0. Similarly, using B_2, the first term on the right hand side of $(*)$ is 0. Hence, L is symmetric.

Lemma 19.6. *If L is symmetric, then $Tx \cdot x \in \mathbb{R}$ for every $x \in \mathcal{D}(T)$.*

Proof: $\overline{Tx \cdot x} = x \cdot Tx = Tx \cdot x$.

Definition 19.7. If T is symmetric, its bounds are defined by

$$m(T) = \inf\{Tx \cdot x : x \in \mathcal{D}(T), \|x\| = 1\}$$

and

$$M(T) = \sup\{Tx \cdot x : x \in \mathcal{D}(T), \|x\| = 1\}.$$

Proposition 19.8. *If T is symmetric and λ is an eigenvalue of T, then $\lambda \in \mathbb{R}$ and $m(T) \leq \lambda \leq M(T)$. Eigenvectors corresponding to distinct eigenvalues are orthogonal.*

Proof: Suppose $Tx = \lambda x$ and $\|x\| = 1$. Then $Tx \cdot x = (\lambda x) \cdot x = \lambda(x \cdot x) = \lambda \in \mathbb{R}$ by Lemma 6 and $m(T) \leq \lambda \leq M(T)$.

If $Tx = \lambda x$ and $Ty = \mu y$ with $\lambda \neq \mu$, then $\lambda(x \cdot y) = (\lambda x) \cdot y = Tx \cdot y = x \cdot Ty = x \cdot (\mu y)$ so $(\lambda - \mu)x \cdot y = 0$ which implies $x \cdot y = 0$.

Theorem 19.9. *If $T : X \to X$ is symmetric and continuous, then*

$$\|T\| = \sup\{|Tx \cdot x| : \|x\| = 1\} = \max\{|m(T)|, |M(T)|\}.$$

Proof: For $\|x\| = 1$ by the Cauchy-Schwarz inequality, $|Tx \cdot x| \leq \|Tx\| \|x\| \leq \|T\|$ so $a = \sup\{|Tx \cdot x| : \|x\| = 1\} \leq \|T\|$. For $x, y \in X$, we have

$$T(x + y) \cdot (x + y) = Tx \cdot x + Ty \cdot y + Tx \cdot y + Ty \cdot x$$
$$= Tx \cdot x + Ty \cdot y + Tx \cdot y + \overline{Tx \cdot y} = Tx \cdot x + Ty \cdot y + 2\mathcal{R}eTx \cdot y$$

and similarly

$$T(x - y) \cdot (x - y) = Tx \cdot x + Ty \cdot y - 2\mathcal{R}eTx \cdot y.$$

Subtraction yields

$$4\mathcal{R}eTx \cdot y = T(x + y) \cdot (x + y) - T(x - y) \cdot (x - y).$$

Since $|Tz \cdot z| \leq a \|z\|^2$ for every $z \in X$, we have

$$(*) \quad |\mathcal{R}eTx \cdot y| \leq (a/4)(\|x + y\|^2 + \|x - y\|^2) = (a/4)(2\|x\|^2 + 2\|y\|^2)$$

by the parallelogram law.

Now suppose $Tx \neq 0$ and set $y = (\|x\| / \|Tx\|)Tx$ in $(*)$ to obtain

$$|Tx \cdot y| = Tx \cdot (\|x\| / \|Tx\|)Tx =$$
$$\mathcal{R}eTx \cdot (\|x\| / \|Tx\|)Tx \leq (a/2)(\|x\|^2 + \|x\|^2) = a\|x\|^2$$

so $\|x\| \|Tx\| \leq a \|x\|^2$ and $\|Tx\| \leq a \|x\|$. This inequality holds also if $Tx = 0$ so $\|T\| \leq a$ and $a = \|T\|$.

The last equality in the statement follows from the definitions.

Exercises.

1. Give a proof of the Hellinger-Toeplitz Theorem using the Closed Graph Theorem.

2. Let c_{00} be given the inner product from l^2 (complex scalars). Show the linear operator $T : c_{00} \to c_{00}$ defined by $T\{t_j\} = \{jt_j\}$ is symmetric but not continuous (thus, completeness in the Hellinger-Toeplitz Theorem is important).

3. A symmetric operator T is *positive* if $Tx \cdot x \geq 0$ for every x in the domain of T. Show that eigenvalues of a positive operator must be non-negative.

4. If H is a Hilbert space and $T \in L(H)$, show T^*T and TT^* are positive.

5. Let $g \in C[a, b]$ (complex values) and define $G : C[a, b] \to C[a, b]$ by $Gf = gf$. When is G symmetric?

Chapter 20

The Spectral Theorem for Compact Symmetric Operators

In this chapter we will establish the spectral theorem for compact symmetric operators. This result is a generalization of a well known result about diagonalizing symmetric real matrices. Namely, if A is a real symmetric $n \times n$ matrix with (real) eigenvalues $\lambda_1, ..., \lambda_n$ and associated orthonormal eigenvectors $x_1, ..., x_n$, then $Ax = \sum_{j=1}^{n} \lambda_j (x \cdot x_j) x_j$ for $x \in \mathbb{R}^n$; that is, the matrix representation of A with respect to $x_1, ..., x_n$ is the diagonal matrix with the $\lambda_1, ..., \lambda_n$ along the diagonal. We establish the analogue of this result for compact symmetric operators.

Let $X \neq \{0\}$ be a complex inner product space and $0 \neq T \in K(X, X)$ symmetric.

Lemma 20.1. *Either $\|T\|$ or $-\|T\|$ is an eigenvalue of T and there exists a corresponding eigenvector x with $\|x\| = 1$ and $|Tx \cdot x| = \|T\|$.*

Proof: By Theorem 19.9, $\|T\| = \max\{|m(T)|, |M(T)|\}$ so there exists $x_n \in X, \|x_n\| = 1$, such that $Tx_n \cdot x_n \to \lambda$, where λ is real and $|\lambda| = \|T\|$ (Theorem 19.9). Now

$$0 \leq \|Tx_n - \lambda x_n\|^2 = \|Tx_n\|^2 - 2\lambda Tx_n \cdot x_n + \lambda^2 \|x_n\|^2$$
$$\leq \|T\|^2 - 2\lambda Tx_n \cdot x_n + \lambda^2 \to 0$$

so $Tx_n - \lambda x_n \to 0$. Since T is compact, $\{Tx_n\}$ has a convergent subsequence $\{Tx_{n_k}\}$. Since $x_{n_k} - (1/\lambda)Tx_{n_k} \to 0$, $\{x_{n_k}\}$ also converges to some $x \in X$. Then $\|x\| = 1$ and $Tx_{n_k} \to Tx$ so $Tx = \lambda x$ and $|Tx \cdot x| = |\lambda| x \cdot x = \|T\|$.

The symmetry of T is important; see Exercise 18.2.

Theorem 20.2. *There exists a possibly finite sequence of non-zero eigenvalues of T, $\{\lambda_k\}$, and a corresponding sequence of eigenvectors $\{x_k\}$ such* *that*

$$(*) \; Tx = \sum_k (Tx \cdot x_k)x_k = \sum_k \lambda_k(x \cdot x_k)x_k.$$

If the sequence $\{\lambda_k\}$ is infinite, $|\lambda_k| \downarrow 0$. Every non-zero eigenvalue of T occurs in the sequence $\{\lambda_k\}$; the eigenmanifold $\ker(\lambda_k - T)$ corresponding to each λ_k is finite dimensional and its dimension is exactly the number of times the eigenvalue λ_k occurs in the sequence $\{\lambda_k\}$.

Proof: Let λ_1 and x_1 be the eigenvalue and eigenvector of T given in Lemma 1. Note $|\lambda_1| = \|T\|$. Let $X_1 = X$ and $X_2 = \{x : x \cdot x_1 = 0\} = \{x_1\}^\perp$. Then X_2 is a closed subspace of X_1 and X_2 is invariant under T since $x \in X_2$ implies $Tx \cdot x_1 = x \cdot Tx_1 = \lambda_1 x \cdot x_1 = 0$. Thus, the restriction of T to X_2 ,T_2, is compact and symmetric. If $T_2 \neq 0$, by Lemma 1 there exist an eigenvalue λ_2 and a corresponding unit eigenvector x_2 with $\|T_2\| = |\lambda_2|$. Then $|\lambda_2| \leq |\lambda_1|$ and $x_1 \cdot x_2 = 0$.

Continuing this construction produces non-zero eigenvalues $\lambda_1, ..., \lambda_k$ and corresponding unit eigenvectors $x_1, ..., x_k$ and closed subspaces $X_1, ..., X_k$ with $X_{j+1} \subset X_j$, $X_{j+1} = \{x_1, ..., x_j\}^\perp$, $x_j \in X_j$, and $|\lambda_j| = \|T_j\|$, where T_j is the restriction of T to X_j. Thus, we have $|\lambda_1| \geq |\lambda_2| \geq ... \geq |\lambda_k|$ and $x_i \cdot x_j = 0$ if $i \neq j$. This construction stops at λ_n, x_n, X_{n+1} iff $T_{n+1} = 0$. In this case $\mathcal{R}T$ lies in the subspace generated by $\{x_1, ..., x_n\}$; for if $x \in X$ and $y_n = x - \sum_{j=1}^n (x \cdot x_j)x_j$, then $y_n \cdot x_j = 0$ $(j = 1, ..., n)$ so $y_n \in X_{n+1}$ which implies that

$$Ty_n = 0 = Tx - \sum_{j=1}^n (x \cdot x_j)Tx_j = Tx - \sum_{j=1}^n \lambda_j(x \cdot x_j)x_j$$

and $Tx \in span\{x_1, ..., x_n\}$. In this case $(*)$ is clearly satisfied. If the construction does not terminate, we obtain infinite sequences $\{\lambda_k\}_{k=1}^\infty, \{x_k\}_{k=1}^\infty$, where $|\lambda_k| \geq |\lambda_{k+1}|, \|x_k\| = 1$ and $\{x_k : k \in \mathbb{N}\}$ is orthonormal.

We claim that $\lambda_k \to 0$. Since $|\lambda_k| \geq |\lambda_{k+1}|$, either $\lambda_k \to 0$ or there exists $\epsilon > 0$ such that $|\lambda_k| \geq \epsilon$ for all k. Suppose the latter holds. Then $\{x_k/\lambda_k\}$ is bounded so $\{T(x_k/\lambda_k)\} = \{x_k\}$ has a convergent subsequence. But, $\{x_k\}$ is orthonormal so $\|x_i - x_j\|^2 = 2$ for $i \neq j$ and $\{x_j\}$ cannot have a convergent subsequence. Hence, $\lambda_k \to 0$.

We next claim that $(*)$ holds when $\{\lambda_k\}$ is infinite. For $x \in X$, set $y_n = x - \sum_{k=1}^n (x \cdot x_k)x_k$. Then $\|y_n\|^2 = \|x\|^2 - \sum_{k=1}^n |x \cdot x_k|^2 \leq \|x\|^2$. Since $y_n \in X_{n+1}$ and $|\lambda_{n+1}| = \|T_{n+1}\|$, $\|Ty_n\| \leq |\lambda_{n+1}| \|y_n\| \leq |\lambda_{n+1}| \|x\|$

so $Ty_n \to 0$. But,

$$Ty_n = Tx - \sum_{k=1}^{n}(x \cdot x_k)Tx_k = Tx - \sum_{k=1}^{n}\lambda_k(x \cdot x_k)x_k$$

so (*) holds.

If $\lambda \neq 0$ is an eigenvalue of T which is not in the sequence $\{\lambda_k\}$, then there is a corresponding unit eigenvector x and $x \cdot x_k = 0$ for all k by Proposition 19.8. By (*) $Tx = 0 = \lambda x$ which is a contradiction. Hence, $\{\lambda_k\}$ exhausts all of the non-zero eigenvalues of T.

Suppose the eigenvalue λ_k occurs N times in the sequence $\{\lambda_k\}$. Then the eigenmanifold $\ker(\lambda_k - T)$ is at least N dimensional. If the dimension of $\ker(\lambda_k - T)$ is greater than N, then there exists x with $\|x\| = 1, Tx = \lambda_k x$ and $x \cdot x_j = 0$ for all j (Proposition 19.8). By (*), $Tx = 0 = \lambda_k x$ which is impossible. Hence, $N = \dim\ker(\lambda_k - T)$.

Note from the proof of Theorem 2, we have a formula for the kth eigenvalue:

$$|\lambda_k| = \max\{|Tx \cdot x| : x \perp span\{x_1, ..., x_k\}, \|x\| = 1\}.$$

From Theorem 2 we can obtain a similar eigenvalue/eigenvector expansion for the resolvent operator of T.

Theorem 20.3. *Let the notation be as in Theorem 2. Let $\lambda \in \mathbb{C}, \lambda \neq 0$ and $\lambda \neq \lambda_k$ for all k. Then $(\lambda - T)^{-1} \in L(X)$ is given by*

$$(\#) \quad (\lambda - T)^{-1}y = \frac{1}{\lambda}y + \frac{1}{\lambda}\sum_k \lambda_k \frac{y \cdot x_k}{\lambda - \lambda_k}x_k = x \text{ for } y \in X.$$

Proof: First suppose the series in $(\#)$ converges to x. Then $(\lambda - T)x = y$ since

$$(\lambda - T)x = y + \sum_k \lambda_k \frac{y \cdot x_k}{\lambda - \lambda_k}x_k - \frac{1}{\lambda}Ty - \frac{1}{\lambda}\sum_k \lambda_k \frac{y \cdot x_k}{\lambda - \lambda_k}Tx_k$$
$$= y + \sum_k \lambda_k \frac{y \cdot x_k}{\lambda - \lambda_k}x_k - \frac{1}{\lambda}\sum_k \lambda_k(y \cdot x_k)x_k - \frac{1}{\lambda}\sum_k \lambda_k^2 \frac{y \cdot x_k}{\lambda - \lambda_k}x_k$$
$$= y + \sum_k (y \cdot x_k)x_k[\frac{\lambda_k}{\lambda - \lambda_k} - \frac{\lambda_k}{\lambda} - \frac{\lambda_k^2}{\lambda(\lambda - \lambda_k)}] = y$$

so $x = (\lambda - T)^{-1}y$.

We show that the series in $(\#)$ converges. Let

$$a = \sup\{\left|\frac{\lambda_k}{\lambda - \lambda_k}\right| : k\} < \infty, b = \sup\{\left|\frac{1}{\lambda - \lambda_k}\right| : k\}, u_n = \sum_{k=1}^{n}\lambda_k \frac{y \cdot x_k}{\lambda - \lambda_k}x_k$$

and $v_n = \sum_{k=1}^{n} \frac{y \cdot x_k}{\lambda - \lambda_k} x_k$. If $m < n$, then

$$\|u_n - u_m\|^2 = \sum_{k=m+1}^{n} \left| \frac{\lambda_k}{\lambda - \lambda_k} \right|^2 |y \cdot x_k|^2 \le a^2 \sum_{k=m+1}^{n} |y \cdot x_k|^2$$

which implies that $\{u_n\}$ is Cauchy since $\sum_{k=1}^{\infty} |y \cdot x_k|^2 \le \|y\|^2$ by the Bessel Inequality. (If X were assumed to be complete, we would be finished.) Now $\|v_n\|^2 = \sum_{k=1}^{n} |y \cdot x_k|^2 / |\lambda - \lambda_k|^2 \le b^2 \sum_{k=1}^{n} |y \cdot x_k|^2 \le b^2 \|y\|^2$ which implies that $\{v_n\}$ is bounded. Note $Tv_n = u_n$ so since T is compact, $\{u_n\}$ must have a convergent subsequence and since it is Cauchy, $\{u_n\}$ must converge.

From (*) of Theorem 2, $\|x\| \le \left|\frac{1}{\lambda}\right| \|y\| + \left|\frac{1}{\lambda}\right| a \|y\|$ so $(\lambda - T)^{-1}$ is continuous and defined on all of X with $\left\|(\lambda - T)^{-1}\right\| \le \left|\frac{1}{\lambda}\right| (1 + a)$.

Corollary 20.4. *Let the notation be as in Theorem 2. Let M be the closure of $span\{x_1, x_2, ...\}$. Then $M^\perp = \ker T$ so $X = M \oplus \ker T$.*

Proof: Let $y \in M^\perp$. Then $y \perp x_k$ for all k so $Ty = \sum_k \lambda_k (y \cdot x_k) x_k = 0$ and $y \in \ker T$ and $M^\perp \subset \ker T$.

Let $y \in \ker T$. Then $y \cdot x_k = \frac{1}{\lambda_k}(y \cdot Tx_k) = \frac{1}{\lambda_k}(Ty \cdot x_k) = 0$ so $y \in M^\perp$ and $\ker T \subset M^\perp$.

Remark 20.5. Thus, if X is complete, then $\{x_1, x_2, ...\}$ is a complete orthonormal set in X iff $\lambda = 0$ is not an eigenvalue of T.

We can also use Theorem 2 to give a spectral representation for arbitrary compact operators between Hilbert spaces.

Theorem 20.6. *Let H_1, H_2 be Hilbert spaces and $T \in K(H_1, H_2)$. Then there exist orthonormal sequences $\{x_k\} \subset H_1$, $\{y_k\} \subset H_2$ and $\lambda_k \in \mathbb{R}$, $\lambda_k \downarrow 0$ such that $Tx = \sum_k \lambda_k (x \cdot x_k) y_k$ for all $x \in H_1$.*

Proof: The operator $T^*T \in K(H_1, H_1)$ and since $T^*Tx \cdot x = Tx \cdot Tx \ge 0$, the non-zero eigenvalue of T^*T are positive; denote these by λ_k^2 and arrange them so that $\lambda_{k+1} \le \lambda_k$. Let $\{x_k\}$ be associated with $\{\lambda_k^2\}$ as in Theorem 2. Set $y_k = (1/\lambda_k)Tx_k$. Then $\{y_k\}$ is orthonormal in H_2 since

$$y_i \cdot y_j = (1/\lambda_i \lambda_j)Tx_i \cdot Tx_j = (1/\lambda_i \lambda_j)T^*Tx_i \cdot x_j = (\lambda_i/\lambda_j)x_i \cdot x_j = 0$$

for $i \ne j$ and $\|y_k\|^2 = (1/\lambda_k^2)T^*Tx_k \cdot x_k = 1$.

We claim the series $\sum_k \lambda_k (x \cdot x_k) y_k$ converges to Tx for $x \in H_1$. First, the series converges in the norm topology of $L(H_1, H_2)$ since

$$\left\| \sum_{k=m}^{n} \lambda_k (x \cdot x_k) y_k \right\|^2 = \sum_{k=m}^{n} |\lambda_k|^2 |x \cdot x_k|^2 \le \lambda_m^2 \sum_{k=m}^{n} |x \cdot x_k|^2 \le \lambda_m^2 \|x\|^2$$

by Bessel's Inequality. Thus, the sum of the series represents a compact operator in $K(H_1, H_2)$. To show that the series converges to the operator T it suffices to establish the equality on a dense subset of H_1. By Corollary 4, $H_1 = \overline{span\{x_k\}} \oplus \ker T^*T$. The equality holds for $x = x_k$ and since $\ker T^*T = \ker T$ the equality holds on H_1.

It follows from Theorem 6 that any compact operator between Hilbert spaces is the norm limit of a sequence of operators with finite dimensional range (Exercise 3).

We now show that the eigenvalue/eigenvector expansions in (*) and (#) of Theorems 2 and 3 are applicable to integral equations.

Let $k \in C([a,b] \times [a,b])$ and let K be the compact operator belonging to $K(C[a,b], C[a,b])$ induced by the kernel k. Assume that $k(s,t) = \overline{k(t,s)}$ for all $s, t \in [a,b]$ so that K is a compact, symmetric operator. Here we are assuming that $C[a,b]$ has the inner product $f \cdot g = \int_a^b f\overline{g}$ and we are using complex scalars; we refer to convergence in the norm induced by the inner product as mean^2 convergence. Thus, a sequence $\{f_k\}$ converges to f mean^2 iff $\int_a^b |f_k - f|^2 \to 0$. Let $\{\lambda_k\}$ be the eigenvalues of K as in Theorem 2 so $|\lambda_1| \geq |\lambda_2| \geq ...$ and let $x_1, x_2, ...$ be the associated eigenvectors.

By Theorem 2

$$(1) \quad Kf = \sum_k \lambda_k (f \cdot x_k) x_k,$$

where the series is mean^2 convergent, and by Theorem 3 if $\lambda \neq 0$ and $\lambda \neq \lambda_k$, the solution of the integral equation $(\lambda - K)f = g$ is given by

$$(2) \quad (\lambda - K)^{-1}g = f = \frac{1}{\lambda}g + \frac{1}{\lambda}\sum_k \lambda_k \frac{g \cdot x_k}{\lambda - \lambda_k} x_k,$$

where the series is mean^2 convergent. We will use the continuity of the kernel k to show the series in (1) and (2) converge uniformly and absolutely.

Theorem 20.7. *(Hilbert/Schmidt) The series in (1) converges uniformly and absolutely.*

Proof: For $t \in [a,b]$ and $f \in C[a,b]$, we have

$$Kf(t) = \int_a^b k(t,s)f(s)ds = \sum_k \lambda_k (f \cdot x_k) x_k(t) \ (mean^2).$$

Now $\lambda_k x_k(t) = Kx_k(t) = \int_a^b k(t,s)x_k(s)ds$, i.e., for fixed t, $\lambda_k x_k(t)$ is the kth Fourier coefficient of the function $k(t, \cdot)$ with respect to $\{x_k\}$. By

Bessel's inequality, $\sum_k |f \cdot x_k|^2 \le \|f\|_2^2$ and $\sum_k |\lambda_k x_k(t)|^2 \le \int_a^b |k(t,s)|^2 ds$. Thus, the series $\sum_k |\lambda_k (f \cdot x_k) x_k(t)|$ converges by the Cauchy-Schwarz Inequality.

Let $M = \max\{|k(t,s)| : t,s \in [a,b]\}$. Then $\int_a^b |k(t,s)|^2 ds \le M^2(b-a)$. By the Cauchy-Schwarz Inequality,

$$\sum_{k=N}^{\infty} |\lambda_k (f \cdot x_k) x_k(t)| \le \left(\sum_{k=N}^{\infty} |f \cdot x_k|^2\right)^{1/2} \left(\sum_{k=N}^{\infty} |\lambda_k x_k(t)|^2\right)^{1/2}$$
$$\le \left(\sum_{k=N}^{\infty} |f \cdot x_k|^2\right)^{1/2} \left(\int_a^b |k(t,s)|^2 ds\right)^{1/2} \le M(b-a)^{1/2} \left(\sum_{k=N}^{\infty} |f \cdot x_k|^2\right)^{1/2}$$

so the series in (1) converges uniformly and absolutely by the Weierstrass M-Test. Note the series converges to $Kf(t)$.

Corollary 20.8. *The series in (2) which gives the solution to the integral equation $(\lambda - K)f = g$ converges uniformly and absolutely for $\lambda \ne 0$ and $\lambda \ne \lambda_k$.*

We can also obtain an eigenvalue/eigenfunction expansion for the kernel k if the kernel is positive. For this we require a lemma.

Lemma 20.9. *If $Kf \cdot f = \int_a^b \int_a^b k(s,t) f(s) \overline{f(t)} ds dt \ge 0$ for all $f \in C[a,b]$, then $k(t,t) \ge 0$ for all $t \in [a,b]$.*

Proof: Suppose $k(t_0, t_0) < 0$ for some $a < t_0 < b$. Then $k(t,t) < 0$ for $t \in [c,d] \times [c,d]$, $a < c < t_0 < d < b$. Pick a continuous function g with support in $[c-\epsilon, d+\epsilon] \subset [a,b]$, $g = 1$ on $[c,d]$ and linear on $[c-\epsilon,c], [d,d+\epsilon]$. Then

$$0 \le \int_a^b \int_a^b k(s,t)g(s)g(t)ds dt$$
$$= \int_c^d \int_c^d k(s,t)ds dt + \int_{c-\epsilon}^{d+\epsilon} \int_{c-\epsilon}^{c} k(s,t)g(s)g(t)ds dt$$
$$+ \int_{c-\epsilon}^{d+\epsilon} \int_{d}^{d+\epsilon} k(s,t)g(s)g(t)ds dt + \int_{c-\epsilon}^{c} \int_{c}^{d} k(s,t)g(s)g(t)ds dt$$
$$+ \int_{d}^{d+\epsilon} \int_{c}^{d} k(s,t)g(s)g(t)ds dt$$
$$\le \int_c^d \int_c^d k(s,t)ds dt + \|k\|_{\infty} [2\epsilon(d-c+2\epsilon) + 2\epsilon(d-c)].$$

The right hand side of this inequality is negative for small ϵ since $\int_c^d \int_c^d k(s,t)ds dt < 0$ giving a contradiction.

Theorem 20.10. *(Mercer) Assume that k is such that*

$$Kf \cdot f = \int_a^b \int_a^b k(s,t) f(s) \overline{f(t)} ds dt \ge 0$$

for all $f \in C[a,b]$. Let $\{\lambda_k\}, \{x_k\}$ be as in Theorem 2. Then $k(s,t) = \sum_k \lambda_k x_k(t) x_k(s)$, where the series converges uniformly and absolutely on $[a,b] \times [a,b]$.

Proof: Since $Kx_k = \lambda_k x_k$, $\lambda_k = Kx_k \cdot x_k \geq 0$, i.e., all the eigenvalues are non-negative. Applying the Cauchy-Schwarz Inequality to $\{\sqrt{\lambda_k} x_k(t)\}$ and $\{\sqrt{\lambda_k} x_k(s)\}$, we obtain

$$(3) \quad \sum_{k=m}^{n} |\lambda_k x_k(t) x_k(s)| \leq (\sum_{k=m}^{n} \lambda_k |x_k(t)|^2)^{1/2} (\sum_{k=m}^{n} \lambda_k |x_k(s)|^2)^{1/2}.$$

We claim

$$(4) \quad \sum_{k=1}^{\infty} \lambda_k |x_k(t)|^2 \leq \max\{k(s,s) : a \leq s \leq b\} = C^2.$$

Let $k_n(t,s) = k(t,s) - \sum_{k=1}^{n} \lambda_k x_k(t) x_k(s)$ so k_n is continuous. Now if f is continuous,

$$\int_a^b \int_a^b k_n(t,s) f(s) \overline{f(t)} ds dt = Kf \cdot f - \sum_{k=1}^{n} \lambda_k |f \cdot x_k|^2 = \sum_{k=n+1}^{\infty} \lambda_k |f \cdot x_k|^2 \geq 0$$

so by Lemma 9, $0 \leq k_n(t,t) = k(t,t) - \sum_{k=1}^{n} \lambda_k |x_k(t)|^2$. Since n is arbitrary, (4) follows.

For fixed t and $\epsilon > 0$, from (3) and (4) there exists N such that $n > m \geq N$ implies $\sum_{k=m}^{n} |\lambda_k x_k(t) x_k(s)| \leq C\epsilon$. Thus, for every t the series $\sum_{k=1}^{\infty} \lambda_k x_k(t) x_k(s)$ converges uniformly and absolutely for $s \in [a,b]$. We now claim that this series converges to $k(t,s)$. Set $h(t,s) = \sum_{k=1}^{\infty} \lambda_k x_k(t) x_k(s)$. If $f \in C[a,b]$ and t is fixed, the uniform convergence of the series implies

$$(5) \quad \int_a^b [k(t,s) - h(t,s)] f(s) ds = Kf(t) - \sum_{k=1}^{\infty} \lambda_k (f \cdot x_k) x_k(t).$$

If $f \in \ker K = \mathcal{R}K^{\perp}$, then since $x_k = (1/\lambda_k) Kx_k \in \mathcal{R}K$ we have $Kf = 0$ and $f \cdot x_k = 0$ so the right hand side of (5) equals 0 in this case. If $f = x_k$ in (5), the right hand side of (5) equals 0. Therefore, for every t, $k(t, \cdot) - h(t, \cdot)$ is perpendicular to $C[a,b]$ by Corollary 4 and is equal to 0. Since this holds for every t, it follows that $k = h$.

In particular, $k(t,t) = \sum_{k=1}^{\infty} \lambda_k |x_k(t)|^2$. The partial sums of this series form an increasing sequence of continuous functions which converges pointwise to $k(t,t)$. By Dini's Theorem ([DeS]11.18), the convergence is uniform for $t \in [a,b]$. Therefore, there exists M such that $n > n \geq M$ implies $\sum_{k=m}^{n} \lambda_k |x_k(t)|^2 < \epsilon^2$ for all t. This along with (3) and (4) imply $\sum_{k=m}^{n} |\lambda_k x_k(t) x_k(s)| < C\epsilon$ for all t, s and the result follows.

The positivity assumption in Mercer's Theorem is important. If Mercer's Theorem holds for the operator in Example 18.7, the function h in this example would be the uniform limit of its Fourier series. But, there are known examples of functions for which this is not the case.

If $A = [a_{ij}]$ is an $n \times n$ matrix with eigenvalues $\lambda_1, ..., \lambda_n$, then the trace formula from linear algebra states that $\sum_{k=1}^{n} \lambda_k = \sum_{k=1}^{n} a_{kk}$. Using Mercer's Theorem, we can obtain an analogue of this result for positive, symmetric, compact operators.

Theorem 20.11. *Assume that k is such that*

$$Kf \cdot f = \int_a^b \int_a^b k(s,t)f(s)\overline{f(t)}dsdt \geq 0$$

for all $f \in C[a,b]$. Let the notation be as in Theorem 2. Then $\sum_k \lambda_k = \int_a^b k(t,t)dt$.

Proof: By Mercer's Theorem, $k(t,t) = \sum_{k=1}^{\infty} \lambda_k |x_k(t)|^2$, where the series converges uniformly for $t \in [a,b]$. Hence, $\int_a^b k(t,t)dt = \sum_{k=1}^{\infty} \lambda_k \|x_k\|_2^2 = \sum_{k=1}^{\infty} \lambda_k$.

Finally, we establish a min-max formula for the eigenvalues of a positive, symmetric, compact operator which does not involve the eigenvectors as in Lemma 1. Results such as this are often employed in numerical approximations.

Theorem 20.12. *Let H be a Hilbert space and T a positive and compact operator with $\lambda_1 \geq \lambda_2 \geq ...$ the eigenvalues of T as in Theorem 2 (recall Exercise 19.3). Then $\lambda_n = \min_{M, \dim M = n-1} \max_{\|x\| \leq 1, x \perp M} Tx \cdot x$.*

Proof: From Lemma 1, $\max\{Tx \cdot x : \|x\| \leq 1, x \perp M\}$ is attained (apply the lemma to PT where P is the projection onto $M^\perp$). We proceed by induction. For $n = 1$, $\lambda_1 = \max\{Tx \cdot x : \|x\| \leq 1\}$ from Lemma 1. Let $x_1, ...$ be the eigenvectors of T corresponding to $\lambda_1, ...$ as in Theorem 2. Given any subspace M of dimension $n - 1$, Lemma 5.27 implies that there exists $x_0 \in span\{x_1, ..., x_n\}$ such that $x_0 \perp M, \|x\| = 1$. Suppose $x_0 = \sum_{j=1}^{n} t_j x_j$. Since $\lambda_j \geq \lambda_n$ for $j = 1, ..., n - 1$,

$$(1) \quad \max_{\|x\| \leq 1, x \perp M} Tx \cdot x \geq Tx_0 \cdot x_0 = \sum_{j=1}^{n} \lambda_j t_j x_j \cdot \sum_{j=1}^{n} t_j x_j$$

$$= \sum_{j=1}^{n} \lambda_j |t_j|^2 \geq \lambda_n \|x_0\|^2 = \lambda_n.$$

But, from the proof of the spectral theorem

$$(2) \quad \lambda_n = \max_{\|x\|=1, x \perp span\{x_1,\ldots,x_{n-1}\}} Tx \cdot x.$$

Since M is an arbitrary subspace of dimension $n-1$, (1) and (2) give the result.

Corollary 20.13. *Let T, S be positive, compact operators on the Hilbert space H. Let $\lambda_1(T) \geq \lambda_2(T) \geq \ldots$ $[\lambda_1(S) \geq \lambda_2(S) \geq \ldots]$ be the eigenvalues of T $[S]$ as in Theorem 2. Then $|\lambda_n(T) - \lambda_n(S)| \leq \|T - S\|$ for all n.*

Proof: If $\|x\| = 1$, then $|Tx \cdot x - Sx \cdot x| = |(T - S)x \cdot x| \leq \|T - S\|$ so $Tx \cdot x \leq Sx \cdot x + \|T - S\|$ and $Sx \cdot x \leq Tx \cdot x + \|T - S\|$. Applying Theorem 12 to these last 2 inequalities gives $\lambda_n(T) \leq \lambda_n(S) + \|T - S\|$ and $\lambda_n(S) \leq \lambda_n(T) + \|T - S\|$ and the result follows.

The inequality in the corollary can be used in approximating eigenvalues. If T_j is a sequence of positive, compact operators converging to the positive, compact operator T, the sequence of the nth eigenvalue of the T_j then converges to the nth eigenvalue of T and can be used to approximate the nth eigenvalue of T.

Remark 20.14. For those readers familiar with the Lebesgue integral, it follows from Example 18.7 and Remark 5 that the orthonormal set $\{e^{ijt} : j \in \mathbb{Z}\}$ is complete in $L^2[-\pi, \pi]$.

Exercises.

1. Let g be a continuous function on $[a, b] \times [a, b]$. Show the function $k(t, s) = \int_a^b \overline{g(x, t)} g(x, s) dx$ satisfies the hypothesis in Mercer's Theorem. Let K be the integral operator induced by k. Show $\sum_k \lambda_k = \int_a^b \int_a^b |g(t, s)|^2 \, ds dt$.

2. Let K be a symmetric, compact operator. Given $z \in X$ show the equation $Kx = z$ has a solution iff $z \in \ker K$ and $\sum_k |z \cdot x_k|^2 / |\lambda_k|^2 < \infty$, where the notation is as in Theorem 2.

3. Show the series in Theorems 2 and 6 converge in operator norm.

Chapter 21

Symmetric Operators with Compact Inverse

Differential operators are not in general compact, but many differential operators have inverses which are compact integral operators. In this chapter we first establish a general theorem for symmetric operators with compact inverse and then apply the theorem to the Sturm-Liouville operators considered in Chapter 19.

Let $X \neq \{0\}$ be an infinite dimensional, complex inner product space and $A : \mathcal{D}(A) \subset X \to X$ be a symmetric linear operator.

Theorem 21.1. *Suppose $T = A^{-1}$ exists and belongs to $K(X,X)$. Let $\{\lambda_k\}, \{x_k\}$ be the eigenvalues and eigenvectors of T as in Theorem 20.2. Set $\mu_k = 1/\lambda_k$. Then the sequence $\{\mu_k\}$ is infinite and $|\mu_k| \to \infty$. For each $x \in \mathcal{D}(A)$, $x = \sum_{k=1}^{\infty}(x \cdot x_k)x_k$. A point $\mu \in \sigma(A)$ iff μ is one of the $\{\mu_k\}$ and $Ax_k = \mu_k x_k$. If $\mu \notin \sigma(A)$, then $(\mu - A)^{-1}y = \sum_{k=1}^{\infty}[(y \cdot x_k)/(\mu - \mu_k)]x_k$ and $(\mu - A)^{-1} \in K(X,X)$.*

Proof: Since A is symmetric, T is symmetric; for if $x, y \in X = \mathcal{D}(T)$ and $Ax_1 = x, Ay_1 = y$, then $x \cdot Ty = Ax_1 \cdot y_1 = x_1 \cdot Ay_1 = Tx \cdot y$. By Corollary 20.4, $X = \overline{span}\{x_k\} \oplus \ker T = \overline{span}\{x_k\}$ so $\{x_k\}$ is infinite since X is infinite dimensional. Hence, $\{\lambda_k\}$ is infinite and, likewise, $\{\mu_k\}$ is infinite with $|\mu_k| \to \infty$ by Theorem 20.2.

If $x \in \mathcal{D}(A)$, then $x = Ty$ for some y so by Theorem 20.2,

$$x = Ty = \sum_k (Ty \cdot x_k)x_k = \sum_k (x \cdot x_k)x_k$$

as advertised.

We claim that if $\mu \neq 0$ and $\mu \neq \mu_k$ for all k, then

$$(\#) \quad (\mu - A)^{-1} = \frac{1}{\mu}T(T - \frac{1}{\mu})^{-1}.$$

Let $x \in \mathcal{D}(A)$ and set $y = (\mu - A)x$. Then $\mu T x - x = Ty$ so $(T - \frac{1}{\mu})x = \frac{1}{\mu}Ty$ and

$$(*) \quad x = \frac{1}{\mu}(T - \frac{1}{\mu})^{-1}Ty = \frac{1}{\mu}T(T - \frac{1}{\mu})^{-1}(\mu - A)x$$

since $\frac{1}{\mu} \in \rho(T)$, $(T - \frac{1}{\mu})^{-1} \in L(X)$ and T and $(T - \frac{1}{\mu})^{-1}$ commute. Suppose $y \in X$ and set $x = \frac{1}{\mu}T(T - \frac{1}{\mu})^{-1}y$ so $x \in \mathcal{D}(A)$ and

$$(**) \quad (\mu - A)x = (\mu - A)\frac{1}{\mu}T(T - \frac{1}{\mu})^{-1}y = (T - \frac{1}{\mu})(T - \frac{1}{\mu})^{-1}y = y.$$

Equations (*) and (**) establish (#) and (#) shows that $(\mu - A)^{-1}$ is compact.

Since $0 \in \rho(A)$ by hypothesis, (#) implies that $\sigma(A) \subset \{\mu_k : k\}$. But, $Tx_k = \lambda_k x_k$ implies $x_k = \lambda_k A x_k$ or $A x_k = \mu_k x_k$ so each μ_k is an eigenvalue of A with associated eigenvector x_k. Thus, $\sigma(A) = \{\mu_k : k\}$.

Finally, from Theorem 20.3,

$$(\frac{1}{\mu} - T)^{-1}y = \mu y + \mu \sum_k \frac{y \cdot x_k}{\mu_k(\frac{1}{\mu} - \frac{1}{\mu_k})}x_k = \mu y + \mu^2 \sum_k \frac{y \cdot x_k}{\mu_k - \mu}x_k$$

so

$$\frac{1}{\mu}T(T - \frac{1}{\mu})^{-1}y = -Ty + \mu \sum_k \frac{y \cdot x_k}{\mu_k(\mu - \mu_k)}x_k$$

$$= -\sum_k \frac{y \cdot x_k}{\mu_k}x_k + \mu \sum_k \frac{y \cdot x_k}{\mu_k(\mu - \mu_k)}x_k = \sum_k \frac{y \cdot x_k}{\mu - \mu_k}x_k$$

$$= (\mu - A)^{-1}y.$$

If X is complete, it then follows from Remark 20.5 that $\{x_k\}$ is a complete orthonormal set.

We apply Theorem 1 to the Sturm-Liouville operator L of Chapter 19. We are considering $X = C[a, b]$ with complex valued functions and the inner product $f \cdot g = \int_a^b f\bar{g}$. Recall $Lf = -(pf')' + qf$, where p, q are real valued functions with $p \in C^1[a, b], p(t) > 0, q \in C[a, b]$, and the domain of L being $\{f \in C^2[a, b] : B_1f = 0, B_2f = 0\}$ with $B_1f = a_1f(a) + a_2f'(a), B_2f = b_1f(b) + b_2f'(b), a_1^2 + a_2^2 > 0, b_1^2 + b_2^2 > 0$.

We describe the eigenvalues of L. The equation $Lf = \lambda f$ reduces to the equation $f'' + \frac{p'}{p}f' + \frac{\lambda - q}{p}f = 0$. Let f_1, f_2 be 2 linearly independent solutions of this equation. We claim that λ is an eigenvalue of L iff

$$(*) \quad \Delta(\lambda) = \det \begin{bmatrix} B_1f_1 & B_1f_2 \\ B_2f_1 & B_2f_2 \end{bmatrix} = 0.$$

Every solution of the differential equation can be written in the form $f = c_1 f_1 + c_2 f_2$. The boundary conditions for f become

$$(**) \quad \begin{aligned} B_1 f &= c_1 B_1 f_1 + c_2 B_1 f_2 = 0 \\ B_2 f &= c_1 B_2 f_1 + c_2 B_2 f_2 = 0. \end{aligned}$$

If λ is an eigenvalue, then $(**)$ has a non-trivial solution (c_1, c_2) so the determinant in $(*)$ is zero. On the other hand, if $\Delta(\lambda) = 0$, then $(**)$ has a unique non-trivial solution (c_1, c_2) and $f = c_1 f_1 + c_2 f_2$ is an eigenvector associated with the eigenvalue λ. This also shows that each eigenvalue is simple or the eigenmanifold associated with each eigenvalue is one dimensional.

We show that if $\lambda = 0$ is not an eigenvalue of L, then L^{-1} is an integral operator whose kernel is called the Green's function of L.

For $j = 1, 2$, let u_j be a non-trivial solution to the differential equation $pu'' + p'u' - qu = 0$ with boundary condition $B_j u = 0$. Since 0 is not an eigenvalue of L, u_1 and u_2 are linearly independent and $B_i u_j \neq 0$ if $i \neq j$. Let W be the Wronskian of u_1, u_2; $W(t) = \det \begin{bmatrix} u_1(t) & u_1'(t) \\ u_2(t) & u_2'(t) \end{bmatrix}$. The general solution of the equation $Lu = v$ is given by

$$u(t) = c_1 u_1(t) + c_2 u_2(t) + \int_a^t \frac{u_1(t)u_2(s) - u_1(s)u_2(t)}{p(a)W(a)} v(s)ds.$$

We want to choose c_1 and c_2 so that $u \in \mathcal{D}(L)$ when v is continuous. If we set $c_2 = 0$ and $c_1 = -\int_a^b [u_2(s)/p(a)W(a)]v(s)ds$, the solution u becomes

$$u(t) = -\int_t^b \frac{u_1(t)u_2(s)}{p(a)W(a)} v(s)ds - \int_a^t \frac{u_2(t)u_1(s)}{p(a)W(a)} v(s)ds = \int_a^b g(t,s)v(s)ds,$$

where $g(t, s) = -u_1(t)u_2(s)/p(a)W(a)$ when $a \leq s \leq t \leq b$ and $g(t, s) = -u_2(t)u_1(s)/p(a)W(a)$ when $a \leq t \leq s \leq b$. The function g is continuous, symmetric, satisfies the boundary conditions defining the domain on L and is called the *Green's function* of L. If G is the integral operator with kernel g, then $G = L^{-1}$.

If 0 is not an eigenvalue of L, Theorems 20.2 and 3 are applicable and give eigenvalue/eigenvector expansions for solutions to both the equations $Lu = v$ and $(\lambda - L)u = v$ when λ is not an eigenvalue of L. It also follows from Theorem 20.7 that these expansions are uniformly and absolutely convergent.

If 0 is not an eigenvalue of L, we can also obtain a pointwise convergent series for the solution of the equation $Lu = v$. By Corollary 20.4, for each fixed t we have $g(t, \cdot) = \sum_k a_k(t) x_k$, where the convergence is mean^2 and

$$a_k(t) = g(t, \cdot) \cdot x_k = \int_a^b g(t, s) x_k(s) ds = G x_k(t) = \lambda_k x_k(t).$$

Observe that if a sequence $\{h_k\}$ converges mean^2 to h, then $\int_a^b |h_k - h| \to 0$ by the Cauchy-Schwarz Inequality so $\int_a^b h_k \to \int_a^b h$. If $v \in C[a, b]$, then $v(\cdot) g(t, \cdot) = \sum_k \lambda_k x_k(t) x_k(\cdot) v(\cdot)$ mean^2 so by the observation just made

$$u(t) = \int_a^b g(t, s) v(s) ds = \sum_k \int_a^b \lambda_k x_k(t) x_k(s) v(s) ds$$

$$= \sum_k \lambda_k x_k(t) \int_a^b x_k(s) v(s) ds = \sum_k \lambda_k x_k(t) c_k,$$

where c_k is the kth Fourier coefficient of v with respect to $\{x_k\}$.

Exercises.

1. If $p = 1, q = 0$, $B_1 f = f(0), B_2 f = f(\pi)$ where $[a, b] = [0, \pi]$, show the eigenvalues of L are $\{k^2\}$ with associated eigenvectors $\{\sin kt\}$.

2. Let $B_1 f = f(a), B_2 f = f(b)$. Show $Lf \cdot f = \int_a^b (p |f'|^2 + q |f|^2)$. If $q \leq 0$, use this to show that 0 is not an eigenvalue and the eigenvalues are positive.

3. Use the formula given to find the Green's function for the operator $Lf = -f'', f(0) = f'(1) = 0$. Find the eigenvalues and associated eigenvectors.

4. Repeat Exercise 3 for $Lf = -f'', f'(0) = f(1) = 0$.

Chapter 22

Bounded Self Adjoint Operators

Let H be a complex Hilbert space. A continuous linear operator $T \in L(H)$ is *self adjoint* if $T = T^*$. Note that if T is self adjoint, then $Tx \cdot y = x \cdot T^* y = x \cdot Ty$ so T is symmetric. Recall the Hellinger-Toeplitz Theorem which asserts that a symmetric linear operator on a Hilbert space is continuous and, therefore, self adjoint.

Proposition 22.1. *The self adjoint operators in $L(H)$ form a (norm) closed real linear subspace containing I.*

Proof: If $T, S \in L(H)$ are self adjoint and $t, s \in \mathbb{R}$, then $(tT + sS)^* = tT^* + sS^* = tT + sS$ so the self adjoint operators form a real linear subspace. If $\{T_j\}$ are self adjoint , $T \in L(H)$ and $\|T_j - T\| \to 0$, then

$$\|T - T^*\| \le \|T - T_j\| + \|T_j^* - T^*\| \to 0$$

so $T = T^*$.

Proposition 22.2. *If $T, S \in L(H)$ are self adjoint, then TS is self adjoint iff $TS = ST$.*

Proof: $\Longrightarrow$: $(TS)^* = S^* T^* = ST = TS$.
$\Longleftarrow$: $(ST)^* = T^* S^* = TS = ST$.

Lemma 22.3. *Let X be a complex inner product space and $T \in L(X)$. If $Tx \cdot x = 0$ for every $x \in X$, then $T = 0$.*

Proof: Let $x, y \in X, s, t \in \mathbb{C}$. Then

$$0 = T(sx + ty) \cdot (sx + ty) = |s|^2 \, Tx \cdot x + |t|^2 \, Ty \cdot y + s\bar{t}Tx \cdot y + \bar{s}tTy \cdot x$$

so

$$(*) \quad 0 = s\bar{t}Tx \cdot y + \bar{s}tTy \cdot x$$

Put $s, t = 1$ in (*) to obtain $Tx \cdot y + Ty \cdot x = 0$; put $s = i, t = 1$ in (*) to obtain $iTx \cdot y - iTy \cdot x = 0$. Thus, $2Tx \cdot y = 0$ and $T = 0$.

The use of complex scalars in Lemma 3 is important; see exercise 1.

Theorem 22.4. $T \in L(H)$ *is self adjoint iff* $Tx \cdot x \in \mathbb{R}$ *for every* $x \in H$.

Proof: If T is self adjoint, then T is symmetric so Lemma 19.6 applies. If $Tx \cdot x \in \mathbb{R}$ for every $x \in H$, then $Tx \cdot x = \overline{Tx \cdot x} = \overline{x \cdot T^*x} = T^*x \cdot x$ so $(T - T^*)x \cdot x = 0$. Hence, $T - T^* = 0$ by Lemma 3.

In the algebra $L(H)$ the self adjoint operators perform a role analogous to the real numbers in the complex numbers as the following result suggests. (See also Exercise 10.)

Theorem 22.5. *If* $T \in L(H)$, *then there exist unique self adjoint operators* A *and* B *such that* $T = A + iB$.

Proof: Set $A = (T + T^*)/2$ and $B = (T - T^*)/2i$.
For uniqueness, suppose $T = A_1 + iB_1$ with A_1, B_1 self adjoint. If $x \in H$, then $Ax \cdot x + iBx \cdot x = A_1x \cdot x + iB_1x \cdot x$ and by Theorem 4 $Ax \cdot x = A_1x \cdot x, Bx \cdot x = B_1x \cdot x$ so $A = A_1, B = B_1$ by Lemma 3.

We can also define an order on the self adjoint operators.

Definition 22.6. Let $T, S \in L(H)$ be self adjoint. Then $T \geq S$ iff $Tx \cdot x \geq Sx \cdot x$ for every $x \in H$. If $T \geq 0$, we say that T is a positive operator. [Note this definition is meaningful by Theorem 4.]

Proposition 22.7. *Let* $U, T, S \in L(H)$ *be self adjoint. If* $T \geq S$, *then* $T + U \geq S + U$ *and* $tT \geq tS$ *for* $t \geq 0$. *This order is a partial order on the self adjoint operators.*

Proof: The first statement is clear. For the last statement, clearly $T \geq T$ and if $T \geq S$ and $S \geq U$, then $T \geq U$. If $T \geq S$ and $S \geq T$, then $(T - S)x \cdot x = 0$ for every x so $T = S$ by Lemma 3.

In the real numbers any positive number has a unique square root; we will establish the analogue of this statement for positive operators later. In

Theorem 22 we will also establish a sequential completeness result for self adjoint operators analogous to that for the real numbers.

Recall that if $M \subset H$, then $M^{\perp} = \{x : x \cdot m = 0 \text{ for every } m \in M\}$. From the Riesz Representation Theorem for H and Proposition 13.14, we have

Proposition 22.8. *If $T \in L(H)$, then $\overline{RT} = (\ker T^*)^{\perp}$.*

Definition 22.9. An operator $T \in L(H)$ is normal if $TT^* = T^*T$.

A self adjoint operator is obviously normal but not conversely (see Exercise 2). If an operator is normal we show that its range and kernel give an orthogonal decomposition for H.

Theorem 22.10. *Let $T \in L(H)$. Then T is normal iff $\|T^*x\| = \|Tx\|$ for every $x \in H$.*

Proof: For $x \in H$, we have $\|Tx\|^2 = Tx \cdot Tx = T^*Tx \cdot x$ and $\|T^*x\|^2 = T^*x \cdot T^*x = TT^*x \cdot x$. If T is normal, these equations imply $\|Tx\|^2 = \|T^*x\|^2$. On the other hand, if $\|T^*x\| = \|Tx\|$, the equations imply $(T^*T - TT^*)x \cdot x = 0$ so $TT^* = T^*T$ by Lemma 3.

Corollary 22.11. *If $T \in L(H)$ is normal, then $\|T^2\| = \|T\|^2$.*

Proof: Replace x by Tx in Theorem 10 to obtain $\|T^*Tx\| = \|T^2x\|$ so $\|T^*T\| = \|T^2\|$. By Proposition 13.22 , $\|T^*T\| = \|T\|^2$ for any $T \in L(H)$.

Theorem 22.12. *If $T \in L(H)$ is normal, then $\overline{RT}$ and $\ker T$ are orthogonal complements so $H = \overline{RT} \oplus \ker T$.*

Proof: By Theorem 10, $\ker T = \ker T^*$ so the result follows from Proposition 8.

We next establish several results which describe the spectrum of a self adjoint operator.

Theorem 22.13. *Let $T \in L(H)$ be normal. Then $\lambda \in \rho(T)$ iff there exists $c > 0$ such that $\|(\lambda - T)x\| \geq c\|x\|$ for all $x \in H$.*

Proof: $\Longrightarrow$: Theorem 2.11.

$\Longleftarrow$: If the condition is satisfied $\lambda - T$ has a continuous inverse by Theorem 2.11 so we must show that the range of $\lambda - T$ is dense. Since T is normal, $\lambda - T$ is normal (Exercise 2) so by Theorem 12, $H = \overline{\mathcal{R}(\lambda - T)} \oplus \ker(\lambda - T) = \overline{\mathcal{R}(\lambda - T)} \oplus \{0\}$ and $\mathcal{R}(\lambda - T)$ is dense.

Corollary 22.14. *Let $T \in L(H)$ be normal. Then $\lambda \in \mathbb{C}$ is in $\sigma(T)$ iff for every $\epsilon > 0$ there exists x, $\|x\| = 1$, such that $\|(\lambda - T)x\| < \epsilon$.*

Recall that a point in the spectrum of a linear operator which satisfies the condition in Corollary 14 is called an approximate eigenvalue (Chapter 17). The set of all approximate eigenvalues is called the approximate spectrum of the operator. From Corollary 14 it follows that the spectrum of a normal operator coincides with its approximate spectrum.

Proposition 22.15. *Let $T \in L(H)$ be normal. Then $R\sigma(T) = \emptyset$.*

Proof: If $\lambda \in \sigma(T)$ is such that $\mathcal{R}(\lambda - T)$ is not dense, then either $\lambda \in P\sigma(T)$ or $\lambda \in R\sigma(T)$. We show that if $\mathcal{R}(\lambda - T)$ is not dense, then $\lambda \in P\sigma(T)$. Now $\overline{\mathcal{R}(\lambda - T)} \neq H$ implies $\mathcal{R}(\lambda - T)^{\perp} \neq \{0\}$ and Theorem 12 implies $\mathcal{R}(\lambda - T)^{\perp} = \ker(\lambda - T) \neq \{0\}$ so $\lambda \in P\sigma(T)$.

Recall the spectral radius of a continuous linear operator T is defined to be $r(T) = \sup\{|\lambda| : \lambda \in \sigma(T)\}$. From Corollary 11 we can obtain a formula for the spectral radius of a normal operator.

Theorem 22.16. *If $T \in L(H)$ is normal, then $r(T) = \|T\|$.*

Proof: By Theorem 16.11, $r(T) = \lim \sqrt[n]{\|T^n\|}$. Since powers of normal operators are normal (Exercise 2), $\|T^n\| = \|T\|^n$ when n is even and the result follows.

Theorem 22.17. *If $T \in L(H)$ is self adjoint, then $\sigma(T) \subset \mathbb{R}$.*

Proof: Let $\lambda = a + bi$ with $b \neq 0$. Let $x \in H$ and set $y = (\lambda - T)x$. Then $y \cdot x = \lambda x \cdot x - Tx \cdot x$ and $x \cdot y = \overline{y \cdot x} = \overline{\lambda} x \cdot x - Tx \cdot x$ since $Tx \cdot x$ is real. Hence, $x \cdot y - y \cdot x = (\overline{\lambda} - \lambda)x \cdot x = -2ibx \cdot x$ and $2|b| \|x\|^2 = |x \cdot y - y \cdot x| \leq 2 \|x\| \|y\|$ by the Cauchy-Schwarz Inequality. Thus, $\|y\| = \|(\lambda - T)x\| \geq |b| \|x\|$ so $\lambda \in \rho(T)$ by Theorem 13.

If T is symmetric, recall the bounds of T are defined to be

$$m(T) = \inf\{Tx \cdot x : \|x\| = 1\}, \quad M(T) = \sup\{Tx \cdot x : \|x\| = 1\}.$$

Theorem 22.18. *If $T \in L(H)$ is self adjoint, then $\sigma(T) \subset [m(T), M(T)]$.*

Proof: Suppose $\lambda \in \mathbb{R}$ is such that $\lambda > M(T)$. Let $\epsilon = \lambda - M(T)$. Then $(\lambda - T)x \cdot x = \lambda x \cdot x - Tx \cdot x \geq \lambda x \cdot x - M(T)x \cdot x = \epsilon \|x\|^2$. Since $(\lambda - T)x \cdot x \leq \|(\lambda - T)x\| \|x\|$, $\|(\lambda - T)x\| \geq \epsilon \|x\|$ and $\lambda \in \rho(T)$ by Theorem 13.

Similarly, if $\lambda < m(T)$, then $\lambda \in \rho(T)$.

Lemma 22.19. *(Generalized Schwarz Inequality) If $T \in L(H)$ is positive, then $|Tx \cdot y|^2 \leq (Tx \cdot x)(Ty \cdot y)$ for $x, y \in H$.*

Proof: Since T is positive, the function $\{x, y\} = Tx \cdot y$ from $H \times H \to \mathbb{C}$ has all of the properties of an inner product except possibly $\{x, x\} = 0$ iff $x = 0$. This property is not required in the proof of the Cauchy-Schwarz Inequality given in Theorem 5.2 so the inequality above follows directly from this proof.

Note that the case when $T = I$ is just the usual Cauchy-Schwarz Inequality.

Theorem 22.20. *If $T \in L(H)$ is self adjoint, then both $m(T)$ and $M(T)$ belong to $\sigma(T)$.*

Proof: Consider the case for $m = m(T)$. For $x \in H$, $(T - m)x \cdot x \geq 0$ so $T - m \geq 0$, i.e., $T - m$ is positive. Apply Lemma 19 to $T - m, x$ and $y = (T - m)x$ to obtain

$$|(T - m)x \cdot (T - m)x|^2 \leq ((T - m)x \cdot x)((T - m)^2 x \cdot (T - m)x)$$

so

$$\|(T - m)x\|^2 \leq ((T - m)x \cdot x) \|T - m\|^3 \|x\|^2$$

and, therefore, $\inf\{\|(T - m)x\| : \|x\| = 1\} = 0$ since $\inf\{(T - m)x \cdot x) : \|x\| = 1\} = 0$ by definition. Hence, $m \in \sigma(T)$ by Corollary 14.

From Theorems 16, 18 and 20, we can now obtain a formula for the spectral radius in terms of the bounds for a self adjoint operator.

Theorem 22.21. *If $T \in L(H)$ is self adjoint, then*

$$\|T\| = r(T) = \max\{|m(T)|, |M(T)|\} = \sup\{|Tx \cdot x| : \|x\| = 1\}.$$

Finally, we establish a sequential completeness property for self adjoint operators which is very analogous to a completeness property for the real numbers.

Theorem 22.22. *Let $\{T_j\}$ be a sequence of self adjoint operators satisfying $T_1 \leq T_2 \leq \dots$ and there exists a self adjoint operator B such that $T_j \leq B$ for all j. Then there exists a self adjoint operator $T \in L(H)$ such that $T_j x \to Tx$ for every $x \in H$ and $T \leq B$.*

Proof: We may assume $0 \leq T_1 \leq T_2 \leq \ldots \leq B$. For $n > m$ set $T_{nm} = T_n - T_m \geq 0$. By the generalized Schwarz Inequality, for $x \in H$, $(T_{nm}x \cdot T_{nm}x)^2 = \|T_{nm}x\|^4 \leq (T_{nm}x \cdot x)(T_{nm}^2 x \cdot T_{nm}x)$. Since $T_{nm} \leq B$, Theorem 21 implies that $\|T_{nm}\| \leq \|B\|$ so

$$(*) \quad \|T_{nm}x\|^4 = \|(T_n - T_m)x\|^4 \leq ((T_n x \cdot x) - (T_m x \cdot x)) \|B\|^3 \|x\|^2.$$

The sequence $\{T_n x \cdot x\}$ is real, increasing and bounded and, therefore, convergent so by $(*)$, the sequence $\{T_n x\}$ is Cauchy and, therefore, converges to some $Tx \in H$. By the Banach-Steinhaus Theorem 8.4, $T \in L(H)$ and $Tx \cdot y = \lim T_n x \cdot y = \lim x \cdot T_n y = x \cdot Ty$ implies that T is self adjoint. Clearly, $T \leq B$.

Theorem 22 can be used to show that any positive operator has a unique square root ([DM],[BN]), but we will give a proof of this fact later by using the spectral theorem for self adjoint operators (26.5).

Exercises.

1. Show Lemma 3 is false for real inner product spaces. [Hint: Use rotations about the origin in the plane.]

2. If T is normal , show T^n is normal for every $n \in \mathbb{N}$ and $\lambda T, \lambda - T$ are normal for every $\lambda \in \mathbb{C}$. Give an example of an operator which is normal but not self adjoint. Give an example of an operator which is not normal.

3. Show the normal operators form a closed subset of $L(H)$.

4. If $T = A + iB \in L(H)$ with A, B self adjoint, show T is normal iff A and B commute (recall Theorem 5). Do the normal operators form a linear subspace of $L(H)$?

5. If p is a real polynomial and T is self adjoint, show $p(T)$ is self adjoint. What about complex polynomials?

6. If T is self adjoint and $-\epsilon I \leq T \leq \epsilon I$, show $\|T\| \leq \epsilon$.

7. Let $\{d_j\} \subset \mathbb{C}$ be bounded and define $D : l^2 \to l^2$ by $D\{t_j\} = \{d_j t_j\}$. Show D is normal and is self adjoint iff the d_j are real. If D is self adjoint, show $m(D) = \inf d_j, M(D) = \sup d_j$.

8. Give an example of a normal operator and a point in its spectrum which is not an eigenvalue.

9. Show that if $\{T_n\}$ is an increasing (decreasing) sequence of self adjoint operators such that $T_n x \cdot x \to Tx \cdot x$, where T is self adjoint, then $T_n x \to Tx$ for every x. [Hint: Consider the proof of Theorem 22.]

10. If $A, B \in L(H)$ are continuous, self adjoint, commuting operators, show $T = A + iB$ is normal. Conversely, show if $T \in L(H)$ is normal, there

exist commuting self adjoint operators A, B such that $T = A + iB$ (recall Theorem 5).

11. Show that if $T \in L(H)$ is self adjoint and T^2 is compact, then T is compact.

12. Show $T \in L(H)$ is normal iff $Tx \cdot Ty = T^*x \cdot T^*y$ for all x, y.

13. Show that if $S \geq 0, T \geq 0$ and $ST = TS$, then $TS \geq 0$.

14. Let $T \in L(H)$. Show T is an isometry iff $\|Tx\| = \|x\|$ for every x iff $T^*T = I$ iff T carries complete orthonormal sets into orthonormal sets.

15. An operator $T \in L(H)$ is *unitary* iff T is an isometry onto H. Show T is unitary iff $T^*T = TT^* = I$ iff T carries complete orthonormal sets to complete orthonormal sets.

16. If $T \in L(H)$ is unitary, show eigenvectors corresponding to distinct eigenvalues are orthogonal.

17. If $T \in L(H)$ is self adjoint, show e^{iT} is unitary (Recall Exercise 2.14.)

Chapter 23

Orthogonal Projections

In this chapter we list some of the basic properties of orthogonal projections for use in the chapter on the Spectral Theorem for self adjoint operators.

Let H be a complex Hilbert space and $P \in L(H)$ be a projection so that $H = \mathcal{R}P \oplus \ker P = \mathcal{R}P \oplus \mathcal{R}(I - P)$.

Definition 23.1. P is an orthogonal projection if $\mathcal{R}P \perp \ker P$, i.e., $\mathcal{R}P^\perp = \ker P$.

We have a characterization of orthogonal projections.

Theorem 23.2. *P is an orthogonal projection iff P is self adjoint.*

Proof: Let $x, y \in H$ with $x = Px + u, y = Py + v, u, v \in \ker P$. Then

$$Px \cdot y = Px \cdot Py + Px \cdot v$$
$$x \cdot Py = Px \cdot Py + u \cdot Py.$$

If P is orthogonal, the equations above imply $Px \cdot y = x \cdot Py$ so P is self adjoint.

If P is self adjoint and $x \in \mathcal{R}P, y \in \ker P$, then $Px = x$ and $Py = 0$ so $x \cdot y = Px \cdot y = x \cdot Py = 0$ and P is orthogonal.

Proposition 23.3. *An orthogonal projection P is positive, $\|P\| \leq 1$ and $0 \leq Px \cdot x \leq 1$ if $\|x\| = 1$. If $P \neq 0$, then $\|P\| = 1$ and if $0 \neq P \neq I$, then $m(P) = 0$ and $M(P) = 1$.*

Proof: P is positive since $Px \cdot x = P^2 x \cdot x = Px \cdot Px = \|Px\|^2 \geq 0$. If $x \in H$ and $x = Px + u$, where $u \in \ker P$, then $\|x\|^2 = \|Px\|^2 + \|u\|^2 \geq \|Px\|^2$ so $\|P\| \leq 1$. If $P \neq 0$, there exists $x \neq 0$ such that $Px = x$ so $\|P\| = 1$.

If $\|x\| = 1$, $0 \leq Px \cdot x = Px \cdot Px = \|Px\|^2 \leq \|P\|^2 \leq 1$.

Since $0 \leq Px \cdot x \leq 1$ for $\|x\| = 1$, $0 \leq m(P) \leq M(P) \leq 1$. If $P \neq 0$, there exists x with $\|x\| = 1$ such that $Px = x$ so $Px \cdot x = x \cdot x = 1$ and $M(P) = 1$. If $P \neq I$, there exists $z \in H$ with $\|z\| = 1$ and $Pz = 0$ so $Pz \cdot z = 0$ and $m(P) = 0$.

Recall we defined a partial order on the self adjoint operators by the condition $S \leq T$ iff $Sx \cdot x \leq Tx \cdot x$ for all x. For orthogonal projections we have the following characterization of this order.

Theorem 23.4. *If P_1, P_2 are orthogonal projections, then $P_1 \leq P_2$ iff $P_2 P_1 = P_1$ (see Exercise 11.4). In this case, $P_2 P_1 = P_1 P_2$ and $P_2 - P_1$ is an orthogonal projection.*

Proof: $\Longleftarrow$: If $P_2 P_1 = P_1$, then $P_1 P_2 x \cdot y = P_2 x \cdot P_1 y = x \cdot P_2 P_1 y = x \cdot P_1 y = P_1 x \cdot y$ for all x, y so $P_1 P_2 = P_1$. Thus, $P_2 - P_1$ is an orthogonal projection and $P_2 - P_1 \geq 0$.

$\Longrightarrow$: If $P_2 \geq P_1$, then $I - P_1 \geq I - P_2$, $I - P_2$ is an orthogonal projection and $(I - P_1)P_1 = 0$ implies

$$0 \leq (I - P_2)P_1 x \cdot (I - P_2)P_1 x = (I - P_2)^2 P_1 x \cdot P_1 x$$
$$= (I - P_2)P_1 x \cdot P_1 x \leq (I - P_1)P_1 x \cdot P_1 x = 0$$

so $(I - P_2)P_1 = 0$ or $P_2 P_1 = P_1$.

Exercises.

1. Two orthogonal projections P_1, P_2 are perpendicular, $P_1 \perp P_2$, iff $P_1 P_2 = 0$. Show $P_1 \perp P_2$ iff $P_2 \perp P_1$ iff $\mathcal{R} P_1 \perp \mathcal{R} P_2$.

2. If P_1, P_2 are orthogonal projections, show $P_1 + P_2$ is an orthogonal projection iff $P_1 \perp P_2$.

3. Let M be a closed subspace of H, P the orthogonal projection onto M and $T \in L(H)$. Show M is invariant under T iff $TP = PTP$.

4. If P is a non-trivial orthogonal projection, show $\sigma(P) = \{0, 1\}$.

Chapter 24

Sesquilinear Functionals

Let X be a complex inner product space.

Definition 24.1. A function $b : X \times X \to \mathbb{C}$ is a sesquilinear functional if $b(\cdot, y)$ is linear for every $y \in X$ and $\overline{b(x, \cdot)}$ is linear for every $x \in X$, i.e., b is linear in the first variable and conjugate linear in the second variable.

For example, the inner product on an inner product space is a sesquilinear functional. More generally, if $B : X \to X$ is linear, then $b(x, y) = Bx \cdot y$ defines a sesquilinear functional.

Definition 24.2. A sesquilinear functional b is bounded if there exists $c \geq 0$ such that $|b(x, y)| \leq c \, \|x\| \, \|y\|$ for every $x, y \in X$. The norm of b is defined to be $\|b\| = \sup\{|b(x, y)| : \|x\| \leq 1, \|y\| \leq 1\}$.

Proposition 24.3. *Let $B : X \to X$ be linear and set $b(x, y) = Bx \cdot y$ for $x, y \in X$. Then B is bounded iff b is bounded. In this case, $\|B\| = \|b\|$.*

Proof: $\Longrightarrow$: By the Cauchy-Schwarz Inequality, $|b(x, y)| \leq \|Bx\| \, \|y\| \leq \|B\| \, \|x\| \, \|y\|$ so b is bounded and $\|b\| \leq \|B\|$.

$\Longleftarrow$: If $x \in X$, $\|Bx\|^2 = Bx \cdot Bx = b(x, Bx) \leq \|b\| \, \|x\| \, \|Bx\|$ so $\|Bx\| \leq \|b\| \, \|x\|$ which implies $\|B\| \leq \|b\|$.

We now show that every bounded sesquilinear functional on a Hilbert space has the form of the functional in Proposition 3.

Theorem 24.4. *If b is a bounded sesquilinear functional on the Hilbert space H, there exists a unique $B \in L(H)$ such that $b(x, y) = Bx \cdot y$ for $x, y \in H$.*

Proof: Fix $x \in H$. Then $\overline{b(x, \cdot)}$ is a bounded linear functional on H so by the Riesz Representation Theorem (5.19) , there is a unique $Bx \in H$

such that $\overline{b(x,y)} = y \cdot Bx$ for all $y \in H$. Since $Bx \cdot y = b(x,y)$, the map B which send x into Bx is linear and is bounded by Proposition 3. Uniqueness is clear.

Definition 24.5. A sesquilinear functional b on X is symmetric if $b(x,y) = \overline{b(y,x)}$ for all $x,y \in X$.

Proposition 24.6. *Let H be a Hilbert space , $B \in L(H)$ and $b(x,y) = Bx \cdot y$ for $x,y \in H$. Then B is self adjoint iff b is symmetric.*

Proof: $\overline{b(y,x)} = \overline{By \cdot x} = x \cdot By$ for $x,y \in H$.

Exercises.

1. Let $f \in C[a,b]$ be real valued. Show $b(x,y) = \int_a^b x(t)\overline{y(t)}f(t)dt$ defines a bounded sesquilinear functional on $C[a,b]$. Find a bounded linear operator so that b has the form of the functional in Theorem 4.

2. Let A, B be bounded linear operators on a Hilbert space H. Show $f(x,y) = Ax \cdot By$ defines a bounded sesquilinear functional on H. Find the bounded linear operator associated with f as in Theorem 4.

Chapter 25

The Spectral Theorem for Bounded Self Adjoint Operators

In this chapter we will establish the spectral theorem for bounded self adjoint operators. To motivate the result we consider the spectral theorem for symmetric, compact operators given in Theorem 20.2. Let T be a compact, symmetric operator with eigenvalues $\{\lambda_k\}$ and associated eigenvectors $\{x_k\}$ so

$$(*) \quad Tx = \sum_k (Tx \cdot x_k)x_k = \sum_k \lambda_k(x \cdot x_k)x_k,$$

where the sum of the series is independent of the order of the terms. Therefore, we may assume that all of the terms where the λ_k are equal are brought together. For each such λ_k let P_k be defined by $P_k x = \sum_{\lambda_i = \lambda_k}(x \cdot x_i)x_i$. Then we have

(i) $P_i = P_k$ if $\lambda_i = \lambda_k$,

(ii) $P_i P_k = 0$ if $\lambda_i \neq \lambda_k$,

(iii) Each P_i is an orthogonal projection.

Then we may write the series $(*)$ above as $Tx = \sum_k' \lambda_k P_k x$, where the prime indicates that we sum over the distinct values of the λ_k. From Corollary 20.4, $X = \overline{span\{x_k\}} \oplus \ker T$ so let P_0 be the orthogonal projection onto $\ker T$. Then $P_0 P_k = P_k P_0 = 0$ if $k \geq 1$. Define a one-parameter family of operators E_λ ($\lambda \in \mathbb{R}$) by $E_\lambda x = \sum_{\lambda_k \leq \lambda}' P_k x$ if $\lambda < 0$ and $E_\lambda x = x - \sum_{\lambda_k > \lambda}' P_k x$. Then each E_λ is an orthogonal projection and $E_\lambda E_\mu = E_\mu E_\lambda = E_\lambda$ for $\lambda \leq \mu$, i.e., $E_\lambda \leq E_\mu$ so the map $\lambda \to E_\lambda$ preserves order, is right continuous in the sense that for each $x \in X$, $\lim_{\lambda \to \mu^+} E_\lambda x = E_\mu x$ and is constant across the intervals determined by the $\{\lambda_k\}$. Formula $(*)$ can now be written as

$$(**) \quad Tx = \int_a^b \lambda dE_\lambda x,$$

where $a < m(T)$, $b \geq M(T)$ and the integral is of the Riemann-Stieltjes type described in Appendix B.

The integral form in (**) suggests that a bounded self adjoint operator may have a similar representation and that is the content of the spectral theorem which we establish in this chapter.

Throughout the remainder of this chapter, let H be a complex Hilbert space and $T \in L(H)$ a self adjoint operator. Recall $\|T\| = \sup\{|m(T)|, |M(T)|\} = r(T)$ (Theorem 22.21). Set $m = m(T), M = M(T)$.

Proposition 25.1. *Let p be a real polynomial. Then $\|p(T)\| \leq \sup\{|p(t)| : m \leq t \leq M\}$.*

Proof: $p(T)$ is self adjoint since p is real so $r(p(T)) = \|p(T)\| = \sup\{|\lambda| : \lambda \in \sigma(p(T))\}$ (Theorem22.21). By the Spectral Mapping Theorem 16.12, $\sigma(p(T)) = p(\sigma(T))$ and since $\sigma(T) \subset [m, M]$ (Theorem 22.18), $\|p(T)\| \leq \sup\{|p(t)| : m \leq t \leq M\}$.

Let $\mathcal{P}$ be the algebra of all real polynomials and equip $\mathcal{P}$ with the sup-norm $\|p\|_\infty = \sup\{|p(t)| : m \leq t \leq M\}$. By Proposition 1, the map $\Psi : \mathcal{P} \to L(H)$ is an algebra homomorphism with norm less than or equal to 1. Let $x, y \in H$ and $p \in \mathcal{P}$. The map $p \to (p(T)x) \cdot y$ from $\mathcal{P}$ into $\mathbb{C}$ is linear and is continuous with norm less than or equal to $\|x\| \|y\|$ since

$$|(p(T)x) \cdot y| \leq \|p(T)\| \|x\| \|y\| \leq \|p\|_\infty \|x\| \|y\|.$$

Since $\mathcal{P}$ is dense in $C[m, M]$ with the sup-norm, this linear map can be extended to a continuous linear functional on $C[m, M]$ with norm less than or equal to $\|x\| \|y\|$. By the Riesz Representation Theorem for $C[m, M]$ described in Appendix C, we have

Proposition 25.2. *There exists a unique function of normalized bounded variation $g(\cdot; x, y) \in NBV[m, M]$ such that*

$$(p(T)x) \cdot y = \int_m^M p(t)dg(t; x, y)$$

for every $p \in \mathcal{P}$ and $Var(g(\cdot; x, y) : [m, M]) \leq \|x\| \|y\|$.

We have the following properties of the function g.

Proposition 25.3. *(i) $g(M; x, y) = x \cdot y$ (ii) $g(t; x_1 + x_2, y) = g(t; x_1, y) + g(t; x_2, y)$ (iii) $g(t; cx, y) = cg(t; x, y)$ (iv) $g(t; x, y) = \overline{g(t; y, x)}$.*

Proof: For (i) put $p = 1$ in Proposition 2.
(ii) through (iv) are similar; for example (iv);

$$\int_m^M t^n dg(t; x, y) = T^n x \cdot y = x \cdot T^n y = \overline{T^n y \cdot x}$$

$$= \overline{\int_m^M t^n dg(t; y, x)} = \int_m^M t^n \overline{dg(t; y, x)}$$

for every $n \geq 0$ so (iv) follows from the Riesz Representation Theorem since the polynomials are dense in $C[m, M]$.

From Proposition 3 and Theorem 24.4 and Proposition 24.6, we have

Proposition 25.4. *For every* $t \in [m, M]$ *there exists a bounded self adjoint operator* $F(t) \in L(H)$ *such that* $F(t)x \cdot y = g(t; x, y)$ *with* $F(m) = 0, F(M) = I$.

We now show that each $F(t)$ is an orthogonal projection. For this we use the property of the Riemann-Stieltjes integral given in Theorem 4 of Appendix B.

Theorem 25.5. *For* $s, t \in [m, M]$, $F(t)F(s) = F(r)$, *where* $r = \min\{s, t\}$. *In particular,* $F(t)^2 = F(t)$, *i.e.,* $F(t)$ *is an orthogonal projection and* $F(t)F(s) = F(s)F(t)$. *[Thus, if* $s \leq t$, *then* $F(s) \leq F(t)$. *]*

Proof: We have

$$\int_m^M t^n d_t\{\int_m^t s^k d_s(F(s)x \cdot y)\} = \int_m^M t^n t^k d(F(t)x \cdot y) = T^{n+k} x \cdot y$$

$$= T^n x \cdot T^k y = \int_m^M t^n d(F(t)x \cdot T^k y)$$

$$= \int_m^M t^n d(T^k F(t)x \cdot y)$$

$$= \int_m^M t^n d\{\int_m^M s^k d_s(F(s)F(t)x \cdot y)\}$$

for every n so

$$\int_m^t s^k d(F(s)x \cdot y) = \int_m^M s^k d_s(F(s)F(t)x \cdot y) = \int_m^M {}^{\bullet} s^k dW(s; x, y),$$

where $W(s; x, y) = F(s)x \cdot y$ if $m \leq s \leq t$ and $W(s; x, y) = F(t)x \cdot y$ if $t \leq s \leq M$. Since $W(\cdot; x, y)$ is a normalized function of bounded variation, we have $W(s; x, y) = F(s)F(t)x \cdot y$ from which the result follows.

We now establish several continuity results for F.

Proposition 25.6. *For $m < t < M$, $x \in X$, $\lim_{s \to t^+} F(s)x = F(t^+)x$ exists and $F(t^+)x = F(t)x$.*

Proof: For $s > t$, $F(s) - F(t)$ is an orthogonal projection by Theorem 23.4 so

$$\|F(s)x - F(t)x\|^2 = (F(s) - F(t))x \cdot x$$
$$= F(s)x \cdot x - F(t)x \cdot x = g(s; x, x) - g(t; x, x) \to 0$$

as $s \to t^+$ since $g(\cdot; x, x)$ is normalized.

Proposition 25.7. *For each $x \in X$, $\lim_{t \to m^+} F(t)x = F(m^+)x$ exists and $F(m^+)$ is an orthogonal projection.*

Proof: Let $m < s < t$. Then

$$\|F(s)x - F(t)x\|^2 = F(s)x \cdot x - F(t)x \cdot x = g(s; x, x) - g(t; x, x) \to 0$$

as $t \to m^+$ since $g(\cdot; x, x)$ is normalized and the result follows.

Similarly, $F(t^-)x = \lim_{s \to t^-} F(s)x$ exists for every t and $x \in H$ and $F(t^-)$ is an orthogonal projection.

Corollary 25.8. *For $m \le t \le M$ and $x \in X$, $F(t)x \cdot x$ is a non-decreasing function of t with $0 \le F(t)x \cdot x = \|F(t)x\|^2 \le \|x\|^2$.*

Proof: For $m \le s < t \le M$, $F(s) \le F(t)$ by Theorem 5 so

$$0 \le F(s)x \cdot x \le F(t)x \cdot x \le F(M)x \cdot x = x \cdot x = \|x\|^2.$$

Proposition 25.9. *For $m \le t \le M$, $F(t)T = TF(t)$, i.e., T commutes with each $F(t)$.*

Proof: For $x, y \in X$,

$$TF(t)x \cdot y = \int_m^M s\, d_s(F(s)F(t)x \cdot y)$$
$$= \int_m^M s\, d_s(F(s)x \cdot F(t)y) = Tx \cdot F(t)y = F(t)Tx \cdot y$$

and the result follows.

From the results above, we have a mapping $F : [m, M] \to L(H)$ with the following properties:

(i) For each $t \in [m, M]$, $F(t)$ is an orthogonal projection with $F(m) = 0, F(M) = 1$.

(ii) If $s \leq t$, then $F(s) \leq F(t)$.

(iii) For each $x, y \in H$, $F(\cdot)x \cdot y \in NBV[m, M]$.

(iv) For each $x \in H$, $F(\cdot)x$ is right continuous on (m, M) and $F(m^+)x$, $F(t^-)x$ exist and are orthogonal projections.

Moreover, we have the following representation for any real polynomial p

$$(\#) \quad p(T)x \cdot y = \int_m^M p(t)dF(t)x \cdot y \ for \ x, y \in H.$$

The function F is called the *resolution of the identity* for the operator T.

We now show that the Riemann-Stieltjes integral in $(\#)$ can be improved to be an integral with respect to the operator norm in $L(H)$.

Theorem 25.10. *If p is a real polynomial, the formula*

$$(*) \quad p(T) = \int_m^M p(t)dF(t)$$

holds, where the integral in (*) is defined as the limit of Riemann-Stieltjes sums which converge in the norm topology of $L(H)$.

Proof: Let $a = t_1 < t_2 < ... < t_n = b$ be a partition of $[m, M]$ and $s_j \in [t_j, t_{j+1}]$ for $j = 1, ..., n-1$. Let

$$S = \sum_{j=1}^{n-1} p(s_j)(F(t_{j+1}) - F(t_j))$$

be the Riemann-Stieltjes sum with respect to this partition and choice of points. Set $\delta_j = \max\{|p(s) - p(t)| : s, t \in [t_j, t_{j+1}]\}$ and $\delta = \max\{\delta_j : 1 \leq j \leq n-1\}$. Then

$$p(T)x \cdot x - Sx \cdot x = \sum_{j=1}^{n-1} \int_{t_j}^{t_{j+1}} (p(t) - p(s_j))dF(t)x \cdot x$$

$$\leq \delta \sum_{j=1}^{n-1} \int_{t_j}^{t_{j+1}} dF(t)x \cdot x \leq \delta x \cdot x$$

from Corollary 8. Therefore, $M(p(T) - S) \leq \delta$. Similarly, $m(p(T) - S) \geq -\delta$. By Theorem 22.21, $\|p(T) - S\| \leq \delta$. Since $\delta \to 0$ as the norm of the partition goes to 0, the result follows.

We give a simple example of the representation above.

Example 25.11. Let $\{d_j\} \subset \mathbb{R}$ and let D be the self adjoint operator $D : l^2 \to l^2$ defined by $Ds = \{d_j s_j\}$ so the operator D is represented by the diagonal matrix with the $\{d_j\}$ down the diagonal. The projection valued function F representing D as above is given by $F(t)x \cdot y = \sum_{d_j \leq t} x_j \overline{y_j}$ so the matrix representing $F(t)$ is the diagonal matrix with 1 in the ith position if $d_i \leq t$ and 0 in the other positions.

Other authors require a normalization for the resolution of the identity in the sense that F is right continuous on the interval $[m, M]$ instead of on $(m, M]$. They replace the resolution F by $E(t) = F(t)$ for $m < t \leq M$ and $E(m) = F(m^+)$; this requires an alteration in the integral representations in (#) and (*) since $E(m)$ may differ from $F(m)$. See Bachman/Narici ([BM]) or Taylor/Lay ([TL]).

There are other approaches to the Spectral Theorem for continuous self adjoint operators. One due to F. Riesz extends the map $\Psi : \mathcal{P} \to L(H)$ defined following Proposition 1 to non-negative, upper semi-continuous functions by using order properties and then to differences of non-negative, upper semi-continuous functions, a class containing the characteristic function of intervals. The extension is then used to construct the projections. See Bachman/Narici ([BN]260) or Riesz/Sz-Nagy ([RN]106,107) for details. Another proof due to Sz-Nagy uses the square root of a positive operator (Theorem 26.5) and the decomposition in Exercise 26.2. See Bachman/Narici ([BN]27) or Riesz/Sz-Nagy ([RN]108) for details.

There is also a Spectral Theorem for normal operators which we outline in Chapter 27.

There is an interesting article on the history of the Spectral Theorem by Steen in [Ste].

Exercises.

1. Show the resolution of the identity in (#) and (*) above is unique.
2. Establish the remarks following Proposition 7.

Chapter 26

An Operational Calculus

In this chapter we use the Spectral Theorem for bounded self adjoint operators to define an operational calculus for bounded self adjoint operators. Throughout this chapter let H be a complex Hilbert space and $T \in L(H)$ a continuous self adjoint operator. We retain the notation from Chapter 25 on the Spectral Theorem. Thus, there exists a function $F : [m, M] \to L(H)$ which has values which are orthogonal projections, is increasing and right continuous on (m, M) with $F(m) = 0$ and $F(M) = I$ and such that

$$(\#) \quad p(T) = \int_m^M p(t)dF(t),$$

for every real polynomial p, where the integral in $(\#)$ is of the Riemann-Stieltjes type converging in the norm topology of $L(H)$.

Recall that in Chapter 25 we defined a map Ψ from the real polynomials $\mathcal{P}$ with the sup norm on $[m, M]$ into $L(H)$ by $p \to p(T)$ and the map Ψ was a continuous, algebraic homomorphism. The map Ψ was then extended to $C_{\mathbb{R}}[m, M]$ and we then obtained from the Riesz Representation Theorem that $\Psi(f)x \cdot y = \int_m^M f(t)dF(t)x \cdot y$ for $x, y \in H$. We now denote the operator $\Psi(f)$ by $F(T)$; note that the operator $f(T)$ is self adjoint being the norm limit of self adjoint polynomial operators. As in the proof of Theorem 25.10 we also have that the integral $\int_m^M f(t)dF(t)$ is a Riemann-Stieltjes type integral which converges in the norm of $L(H)$ and we write $f(T) = \int_m^M f(t)dF(t)$. Thus we have a map $f \to f(T) = \int_m^M f(t)dF(t)$ from $C_{\mathbb{R}}[m, M] \to L(H)$, where each $f(T)$ is self adjoint. This map is often referred to as an *operational calculus*. We now establish the basic properties of this map.

Theorem 26.1. *(i)* $(f + g)(T) = f(T) + g(T)$, *(ii)* $(cf)(T) = cf(T)$, *(iii)* $(fg)(T) = f(T)g(T)$, *(iv)* $\|f(T)\| \leq \sup\{|f(t)| : m \leq t \leq M\} = \|f\|_{\infty}$, *(v)* *if $B \in L(H)$ commutes with each $F(t)$, then B commutes with $f(T)$, (vi) $f(T)$ is positive if $f \geq 0$, (vii) $\|f(T)x\|^2 = \int_m^M |f(t)|^2 \, dF(t)x \cdot x$.*

Proof: (i) and (ii) are clear.

For (iii) we use the Riemann-Stieltjes approximating sums. Let $m = t_1 < t_2 < ... < t_n = M$ and $s_j \in [t_j, t_{j+1}]$ for $j = 1, ..., n-1$. If $\sum_{j=1}^{n-1} f(s_j)(F(t_{j+1}) - F(t_j))$ and $\sum_{j=1}^{n-1} g(s_j)(F(t_{j+1}) - F(t_j))$ are the corresponding Riemann-Stieltjes sums, then using the relationships $(F(t_{j+1}) - F(t_j))(F(t_{k+1}) - F(t_k)) = (F(t_{j+1}) - F(t_j))$ if $j = k$ and $(F(t_{j+1}) - F(t_j))(F(t_{k+1}) - F(t_k)) = 0$ if $j \neq k$ and multiplying the approximating sums gives $\sum_{j=1}^{n-1} f(s_j)g(s_j)(F(t_{j+1}) - F(t_j))$, which is an approximating sum for fg. Now taking limits gives the result.

For (iv) we have

$$|f(T)x \cdot x| = \left| \int_m^M f(t)dF(t)x \cdot x \right| \leq \|f\|_\infty \int_m^M dF(t)x \cdot x = \|f\|_\infty \, x \cdot x$$

and the result follows from Theorem 22.21.

We can establish (v) by using approximating sums. Let S be the approximating sum for $f(T)$ above. Then we have $BS = SB$ so passing to the limit gives (v).

(vi) is clear. For (vii), from (iii), we have

$$\|f(T)x\|^2 = f(T)x \cdot f(T)x = f(T)f(T)x \cdot x = f^2(T)x \cdot x = \int_m^M f^2(t)dF(t)x \cdot x.$$

Note that (iii) means that $\int f dF \int g dF = \int fg dF$; i.e., the integral of the product is the product of the integrals!

We next establish a commutativity result for T.

Lemma 26.2. *For every $t \in [m, M]$, there exists a sequence of polynomials $\{p_n\}$ such that $p_n(T)x \to F(t)x$ for $x \in H$.*

Proof: For $t = m$ or $t = M$, the result is clear. Assume $m < t < M$. For small h define $f_h(s) = 1$ if $m \leq s \leq t$, $f_h(s) = 1 - (s - t)/h$ if $t \leq s \leq t + h$ and $f_h(s) = 0$ if $t + h \leq s \leq M$. Then

$$f_h(T)x \cdot x = \int_m^t dF(s)x \cdot x + \int_t^{t+h} f_h(s)dF(s)x \cdot x$$

$$= F(t)x \cdot x + \int_t^{t+h} f_h(s)dF(s)x \cdot x$$

since $F(s)x \cdot x$ is increasing and $F(m)x \cdot x = 0$. As $h \to 0$, $\left| \int_t^{t+h} f_h(s) dF(s)x \cdot x \right| \le F(t+h)x \cdot x - F(t)x \cdot x \to 0$ by right continuity of $F(s)x$. Thus, $f_h(T)x \cdot x \to F(t)x \cdot x$. By Exercise 22.9 $f_h(T)x \to F(T)x$ [note $f_h(T) \downarrow$]. For each n pick a polynomial p_n such that $\|f_{1/n} - p_n\|_\infty < 1/n$. Then $\|f_{1/n}(T) - p_n(T)\| \le \|f_{1/n} - p_n\|_\infty < 1/n$ by Theorem 1. Hence,

$$\|p_n(T)x - F(t)x\| \le \|f_{1/n}(T)x - p_n(T)x\| + \|f_{1/n}(T)x - F(t)x\|$$
$$< 1/n + \|f_{1/n}(T)x - F(t)x\| \to 0$$

as $n \to \infty$.

Theorem 26.3. *If $B \in L(H)$ commutes with T, then B commutes with each $F(t)$.*

Proof: For every polynomial p, B commutes with $p(T)$. Pick a sequence of polynomials $\{p_n\}$ as in Lemma 2. Then $Bp_n(T)x = p_n(T)Bx \to BF(t)x = F(t)Bx$ for every x and the result follows.

Corollary 26.4. *If $B \in L(H)$ commutes with T, then B commutes with $f(T)$ for every continuous function f.*

Proof: This follows from Theorems 1 and 3.

As an application of the operational calculus, we establish the existence of the square root for a positive operator.

Theorem 26.5. *If T is a positive operator, then T has a unique positive square root denoted by $\sqrt{T}$ (that is, $(\sqrt{T})^2 = T$). Moreover, if $B \in L(H)$ commutes with T, then B commutes with $\sqrt{T}$.*

Proof: Let $f(t) = \sqrt{t}$ for $t \ge 0$. Since $[m, M] \subset [0, \infty)$, $f(T)$ is well defined and $f(T)^2 = T$ so we define $f(T) = \sqrt{T}$.

That $\sqrt{T}$ commutes with every operator which commutes with T follows from Corollary 4.

For uniqueness suppose C is positive and $C^2 = T$. Set $B = \sqrt{T}$. Then

$$(*) \quad \left\|\sqrt{C}x\right\|^2 + \left\|\sqrt{B}x\right\|^2 = Cx \cdot x + Bx \cdot x = (C + B)x \cdot x$$

for $x \in H$. Since $TC = C^2 C = CC^2 = CT$, C commutes with B by what is noted above. For $y \in H$ put $x = (B - C)y$ in (*) to obtain

$$\left\|\sqrt{C}x\right\|^2 + \left\|\sqrt{B}x\right\|^2 = (C+B)x \cdot x = (C+B)(B-C)y \cdot x = (B^2 - C^2)y \cdot x = 0$$

so $\sqrt{C}x = 0$ and $\sqrt{B}x = 0$. Hence, $Cx = \sqrt{C}\sqrt{C}x = 0$ and $Bx = 0$ and $\|(B - C)y\|^2 = (B - C)^2 y \cdot y = (B - C)x \cdot y = 0$ so $(B - C)y = 0$ and $B = C$.

In Theorem 22.5 we noted that the self adjoint operators could be viewed as playing the role of the real numbers in $L(H)$ or from Exercise 22.10 the normal operators could be viewed as playing the role of the complex numbers in $L(H)$. Every complex number z has a polar decomposition of the form $z = re^{i\theta}$ as well as the representation $z = a + ib$. We use the square root to show that every normal operator also has such a polar decomposition. See also Exercise 2 for another decomposition for continuous self adjoint operators into positive and negative parts.

Theorem 26.6. (*Polar Decomposition*) *Let* $T \in L(H)$ *be normal. There exist a positive operator* P *and a linear isometry* U *such that* $T = UP$. *Moreover, any operator in* $L(H)$ *which commutes with* T *commutes with* U *and* P.

Proof: Note that TT^* is positive and set $P = \sqrt{TT^*}$. Note

$$(*) \quad Px \cdot Px = \|Px\|^2 = x \cdot P^2 x = x \cdot TT^* x = Tx \cdot Tx = \|Tx\|^2.$$

Thus, $\ker P = \ker T$ and $H = \overline{\mathcal{R}T} \oplus \ker T = \overline{\mathcal{R}P} \oplus \ker P$ from Theorem 22.12. Hence, $\mathcal{R}T$ and $\mathcal{R}P$ have the same closure. Now define a linear operator U on $\mathcal{R}P$ by $U(Px) = Tx$; note that U is well defined since if $Px = Py$, then $Tx = Ty$ by $(*)$ applied to $x - y$ and U is an isometry on $\mathcal{R}P$ by $(*)$. We may extend U to $\overline{\mathcal{R}P} = \overline{\mathcal{R}T}$ by continuity and U will still be a linear isometry on $\overline{\mathcal{R}P}$. We may extend U to be an isometry on H by letting U be the identity on $\overline{\mathcal{R}P}^{\perp} = \overline{\mathcal{R}T}^{\perp} = \ker P = \ker T$. We claim that $T = UP$. If $x \in \mathcal{R}P$, $UPx = Tx$ by definition; if $x \in \ker P = \ker T$, $Tx = UPx = 0$.

Suppose $B \in L(H)$ commutes with T. Then $BTx = BUPx = TBx = UPBx = UBPx$, since B commutes with P , which implies $BU = UB$ on $\mathcal{R}P$ and, therefore, on $\overline{\mathcal{R}P}$. On the other hand, if $x \in \overline{\mathcal{R}P}^{\perp} = \overline{\mathcal{R}T}^{\perp} = \ker T = \ker P$, then $Bx \in \overline{\mathcal{R}P}^{\perp}$ since $PBx = BPx = 0$ so $BUx = BIx = UBx$ by the definition of U on $\overline{\mathcal{R}P}^{\perp}$. Hence, B commutes with U; B commutes with P by Theorem 5.

In the next results we show that the operational calculus and the projection valued function F can be used to characterize the various points in the spectrum of T. Recall the spectrum of T lies inside the interval $[m, M]$.

Theorem 26.7. *The point $\lambda \in [m, M]$ belongs to $\rho(T)$ iff there exists $\epsilon > 0$ such that $F(t)$ is constant on the interval $[\lambda - \epsilon, \lambda + \epsilon]$.*

Proof: Consider the "if" statement first. Set $f(s) = \lambda - s$ and let $g(s) = 1/(\lambda-s)$ outside $[\lambda-\epsilon, \lambda+\epsilon]$ and define $g(s)$ arbitrarily on $[\lambda-\epsilon, \lambda+\epsilon]$ except that g must be continuous. Then $f(s)g(s) = 1$ except on $[\lambda-\epsilon, \lambda+\epsilon]$. Since F is constant on $[\lambda - \epsilon, \lambda + \epsilon]$, $f(T)g(T) = \int_m^M f(s)g(s)dF(s) = \int_m^M dF(s) = I$. Therefore, $g(T) = f(T)^{-1} = (\lambda - T)^{-1}$.

For the other statement, we may assume $m < \lambda < M$ (Theorem 22.20). Suppose for every small $\epsilon > 0$ there exist $t_1 < t_2$ in $[\lambda - \epsilon, \lambda + \epsilon]$ such that $F(t_1) \neq F(t_2)$. Since $F(t_1) = F(t_2)F(t_1)$, the range M_1 of $F(t_1)$ is properly contained in the range M_2 of $F(t_2)$ so there exists $y \in M_2 \backslash M_1$ such that $y \perp M_1$. Then $F(t_2)y = y$ and $F(t_1)y = 0$. If $t \leq t_1$, $F(t)y = F(t)F(t_1)y = 0$ and if $t_2 \leq t$, then $F(t)y = F(t)F(t_2)y = F(t_2)y = y$. Since $F(t)y$ is constant if $t \leq t_1$ or $t \geq t_2$, Theorem 25.10 implies

$$\|(\lambda - T)y\|^2 = \int_{t_1}^{t_2} (\lambda - t)^2 dF(t)y \cdot y \leq \epsilon^2 \|y\|^2$$

since $F(t)y \cdot y = \|F(t)y\|^2 \leq \|y\|^2$ and $F(t)y \cdot y \uparrow$. Therefore, $\inf_{\|y\|=1} \|(\lambda - T)y\| = 0$ which implies that $\lambda \in \sigma(T)$ (Corollary 22.14).

Remark 26.8. If $\lambda \in \rho(T)$, then by the proof of the theorem above the resolvent operator $R_\lambda = (\lambda - T)^{-1}$ is given by $R_\lambda = \int_m^M \frac{1}{\lambda-t}dF(t)$; the values of $F(t)$ are constant on an interval about λ so the singularity at λ causes no difficulties.

Next, we have a characterization of the point spectrum or the eigenvalues of T.

Theorem 26.9. *A point λ lies in $P\sigma(T)$ iff $F(\lambda) \neq F(\lambda^-)$. If $\lambda \in P\sigma(T)$, the eigenspace of λ is the range of the projection $F(\lambda) - F(\lambda^-)$.*

Proof: Consider the "if" statement first. If $t \leq s$, then $F(t)F(s) = F(t)$ so $F(t)F(\lambda^-) = F(\lambda^-)$ if $\lambda \leq t$ and $F(t)F(\lambda^-) = F(t)$ if $t < \lambda$. Therefore, $F(\lambda^-)$ and $F(\lambda) - F(\lambda^-)$ are projections (remark following Proposition 25.7). Suppose $F(\lambda) \neq F(\lambda^-)$. Let $y = (F(\lambda)-F(\lambda^-))x$ with $y \neq 0$. Then $F(t)y = 0$ if $t < \lambda$ and $F(t)y = y$ if $\lambda \leq t$. From Theorem 25.10

$$\|(\lambda - T)y\|^2 = \int_m^M (\lambda - t)^2 dF(t)y \cdot y = \int_m^\lambda (\lambda - t)^2 dF(t)y \cdot y = 0$$

since $(\lambda - t) = 0$ at $t = \lambda$. Hence, $\lambda \in P\sigma(T)$ and y is an associated eigenvector. Note that y belongs to the range of $F(\lambda)-F(\lambda^-)$ so $\mathcal{R}(F(\lambda)-F(\lambda^-))$ is contained in the eigenspace of λ.

For the converse, suppose $\lambda \in P\sigma(T)$, $(\lambda - T)y = 0$ with $y \neq 0$. We extend the domain of F by setting $F(t) = 0$ if $t < m$ and $F(t) = I$ if $t > M$. Then $\int_a^b (\lambda - t)^2 dF(t)y \cdot y = 0$, where $a < m$ and $b > M$ can be chosen arbitrarily. Choose $\epsilon > 0$, a and b such that $a < \lambda - \epsilon$ and $\lambda + \epsilon < b$. Then

$$0 \leq \epsilon^2(\|F(\lambda - \epsilon)y\|^2 - \|F(a)y\|^2) = \epsilon^2 \|F(\lambda - \epsilon)y\|^2$$

$$\leq \int_a^{\lambda-\epsilon} (\lambda - t)^2 dF(t)y \cdot y \leq \int_a^b (\lambda - t)^2 dF(t)y \cdot y = 0$$

so $F(\lambda - \epsilon)y = 0$ and $F(\lambda^-)y = 0$. Similarly, $\int_{\lambda+\epsilon}^b (\lambda - t)^2 dF(t)y \cdot y = 0$ and $0 = \epsilon^2(\|F(b)y\|^2 - \|F(\lambda + \epsilon)y\|^2) = \epsilon^2(\|y\|^2 - \|F(\lambda + \epsilon)y\|^2)$ so $F(\lambda + \epsilon)y = y$ and $F(\lambda)y = y$ by right continuity. Hence, $(F(\lambda) - F(\lambda^-))y = y$ so $F(\lambda) \neq F(\lambda^-)$ and $y \in \mathcal{R}(F(\lambda) - F(\lambda^-))$ so any non-zero eigenvector associated with λ belongs to $\mathcal{R}(F(\lambda) - F(\lambda^-))$.

From Theorems 7, 9 and Proposition 22.15, we obtain a characterization of the continuous spectrum of T.

Theorem 26.10. *$C\sigma(T)$ consists of those points λ such that $F(\lambda) = F(\lambda^-)$ but $F(t)$ is not constant on any interval containing λ.*

Finally, we extend the Spectral Mapping Theorem given in Theorem 16.12. For this we require some additional results which are of interest in themselves.

Lemma 26.11. *If p is a real polynomial, then $\|p(T)\| = \max\{|p(t)| : t \in \sigma(T)\}$.*

Proof:

$$(*) \quad \|p(T)\|^2 = \sup\{p(T)x \cdot p(T)x : \|x\| = 1\} = \sup\{p^2(T)x \cdot x : \|x\| = 1\}.$$

Now $p^2(T)$ is a positive operator so $M(p^2(T)) = \sup\{\lambda : \lambda \in \sigma(p^2(T))\}$. From the Spectral Mapping Theorem 16.12

$$(**) \quad M(p^2(T)) = \sup\{\lambda : \lambda \in \sigma(p^2(T))\}$$
$$= \sup\{p^2(t) : t \in \sigma(T)\} = (\sup\{|p(t)| : t \in \sigma(T)\})^2$$

and the result follows from (*) and (**).

Theorem 26.12. *If f is a real valued continuous function on $[m, M]$, then $\|f(T)\| = \sup\{|f(t)| : t \in \sigma(T)\}$.*

Proof: Choose a sequence of polynomials $\{p_n\}$ which converge uniformly to f on $[m, M]$. By Lemma 11 and Theorem 1, $\|p_n(T)\| = \sup\{|p(t)| : t \in \sigma(T)\} \to \|f(T)\| = \sup\{|f(t)| : t \in \sigma(T)\}$.

Corollary 26.13. *If f_1, f_2 are continuous real valued functions on $[m, M]$ such that $f_1 = f_2$ on $\sigma(T)$, then $f_1(T) = f_2(T)$.*

Proof: $\|f_1(T) - f_2(T)\| = \sup\{|f_1(t) - f_2(t)| : t \in \sigma(T)\}$ by Theorem 12.

Remark 26.14. Thus, if f is a continuous function defined on $\sigma(T)$ and is extended to a continuous function f^*, then the operator $f^*(T)$ is independent of the particular extension.

We now have the machinery in place to extend the Spectral Mapping Theorem.

Theorem 26.15. *(Spectral Mapping Theorem) If f is a continuous real valued function of $[m, M]$, then $f(\sigma(T)) = \sigma(f(T))$.*

Proof: Suppose $\lambda \notin f(\sigma(T))$. Then $f - \lambda \neq 0$ on $\sigma(T)$. Therefore, $g(t) = 1/(\lambda - t) \neq 0$ is continuous on $\sigma(T)$. Extend g continuously to $[m, M]$. Then $(f(t) - \lambda)g(t) = 1$ on $\sigma(T)$ so by Remark 14 $(f(T) - \lambda I)g(T) = I$ and $g(T) = (f(T) - \lambda I)^{-1}$ so $\lambda \notin \sigma(f(T))$. Therefore, $\sigma(f(T)) \subset f(\sigma(T))$.

Conversely, suppose $\lambda \in \sigma(T)$. We show $f(T) - f(\lambda)I$ is not invertible. Choose a sequence of polynomials $\{p_n\}$ such that $p_n \to f$ uniformly on $[m, M]$. Then $p_n(T) - p_n(\lambda)I \to f(T) - f(\lambda)I$ in norm. By the Spectral Mapping Theorem 16.12, $p_n(\lambda) \in \sigma(p_n(T))$ so $p_n(T) - p_n(\lambda)I$ is not invertible which implies that $f(T) - f(\lambda)I$ is not invertible (Corollary 2.18). Therefore, $f(\sigma(T)) \subset \sigma(f(T))$.

Exercises.

1. If T is a compact, symmetric, positive operator, give a series expansion for $\sqrt{T}$ as in (*) of Theorem 20.2.

2. If $T \in L(H)$ is self adjoint, show there exist positive operators T^+, T^- such that $T = T^+ - T^-$ and T^+, T^- commute with any operator in $L(H)$ which commutes with T. Show $T^+T^- = 0$, $-T^- \leq T \leq T^+$ and $T = T^+$ if $T \geq 0$. [Hint: Use the function $f(t) = (|t| + t)/2$.]

3. Show that if $T \geq 0$ is invertible, then $\sqrt{T}$ is invertible.

4. If $f \in C[m, M]$ is such that $f \neq 0$ on $\sigma(T)$, show $f(T)$ is invertible. (Hint: Use Remark 14 with the function $1/f$.)

5. Show that if $f(T)$ is invertible, then $f \neq 0$ on $\sigma(T)$. (Hint: Suppose $f(t_0) = 0$ for some $t_0 \in \sigma(T)$. Use Theorem 10 to find $t_n \to t_0^+$ such that $P_n = F(t_n) - F(t_0) \neq 0$. Pick $x_n \in \mathcal{R}P_n$ such that $\|x_n\| = 1$. Show $f(T)x_n = f(T)P_n x_n = \int_{t_0}^{t_n} f(t)dF(t)x_n$ and $\|f(T)x_n\|^2 = \int_{t_0}^{t_n} f(t)^2 dF(t)x_n \cdot x_n \leq \|x_n\|^2 \max\{|f(t)|^2 : t_0 \leq t \leq t_n\} \to 0$ and use 2.11.)

6. If $T \in L(H)$ is compact and self adjoint, is $f(T)$ compact for $f \in C[m, M]$?

Chapter 27

The Spectral Theorem for Normal Operators

In this chapter we sketch a proof of a spectral theorem for normal operators in the spirit of the spectral theorem for bounded, self adjoint operators given in Chapter 25. Complete details may be found in Riesz-Nagy ([RN]) or Bachman/Narici ([BN]).

Let H be a complex Hilbert space and $T \in L(H)$ a normal operator. Then there exist unique bounded, self adjoint operators $A, B \in L(H)$ such that $T = A + iB$ and A and B commute (Theorem 22.5 and Exercise 22.4). Let E, F be the resolutions of the identity for A, B, respectively. Extend E [F] to $\mathbb{R}$ by setting $E(s) = 0$ if $s < m(A)$ and $E(s) = I$ if $s > M(A)$ [$F(t) = 0$ if $t < m(B)$ and $F(t) = I$ if $t > M(B)$], where $m(A), M(A)$ are the bounds for A (19.7). Then $A = \int_{-\infty}^{\infty} s\, dE(s)$, $I = \int_{-\infty}^{\infty} dE(s)$, $B = \int_{-\infty}^{\infty} t\, dF(t)$, $I = \int_{-\infty}^{\infty} dF(t)$. Since A and B commute, each $E(s), F(t)$ commute (Theorems 26.1 and 26.3). Thus, we formally have

$$(*) \quad T = A + iB = \int_{-\infty}^{\infty} s\, dE(s) \int_{-\infty}^{\infty} dF(t) + i \int_{-\infty}^{\infty} dE(s) \int_{-\infty}^{\infty} t\, dF(t)$$

$$= \int_{-\infty}^{\infty} \int_{-\infty}^{\infty} (s + it)\, dE(s)\, dF(t).$$

The iterated integral in $(*)$ suggests the possibility of writing T as a "double integral" over the complex plane and obtaining a representation like that for a bounded, self adjoint operator given in Chapter 25.

For a half-closed rectangle $\delta = (x_1, x_2] \times (y_1, y_2]$, set

$$G(\delta) = (E(x_2) - E(x_1))(F(y_2) - F(y_1)).$$

The function G defined on the half-closed rectangles has the following properties:

(1) Since each $E(s), F(t)$ commute, each $G(\delta)$ is an orthogonal projection.

159

(2) G is additive in the sense that if δ_1, δ_2 are half-closed rectangles whose union δ is a half-closed rectangle, then $G(\delta) = G(\delta_1) + G(\delta_2)$.

(3) If we set $G(\emptyset) = 0$, then $G(\delta_1)G(\delta_2) = G(\delta_1 \cap \delta_2)$ for all half-closed rectangles δ_1, δ_2.

(4) If δ_1, δ_2 are disjoint, then $G(\delta_1) \perp G(\delta_2)$.

(5) Set $\Delta = [m(A), M(A)] \times [m(B), M(B)]$ and let δ be a half-closed rectangle. If $\delta \supset \Delta$, then $G(\delta) = I$ and if $\delta \cap \Delta = \emptyset$ then $G(\delta) = 0$.

Consider decomposing the complex plane into pairwise disjoint half-closed intervals $\delta_{hk} = (x_{h-1}, x_h] \times (y_{k-1}, y_k]$, $h, k \in \mathbb{Z}$. From the properties of G listed above, the family $\{G(\delta_{hk})\}$ defines a decomposition of H into mutually orthogonal subspaces (note from (5) there are only a finite number of non-zero terms in the decomposition). We consider defining a double integral over the plane with respect to G. Set $E_h = E(x_h) - E(x_{h-1})$, $F_k = F(y_k) - F(y_{k-1})$ so $G(\delta_{hk}) = E_h F_k$. Let $z_{hk} = x_{hk} + iy_{hk} \in \delta_{hk}$ and form the Riemann type (finite) sum with respect to G by

$$\sum_{h,k} z_{hk} G(\delta_{hk}) = \sum_{h,k} z_{hk} E_h F_k$$

to approximate the iterated integral representing T in (*). Suppose $|x_h - x_{h-1}| < \epsilon$ so $|x_{hk} - x_k| \le \epsilon$ for every k or $x_h - \epsilon \le x_{hk} \le x_h + \epsilon$ for every k. Then

(#) $\displaystyle\sum_{h,k} (x_h - \epsilon) E_h F_k \le \sum_{h,k} x_{hk} E_h F_k \le \sum_{h,k} (x_h + \epsilon) E_h F_k$ (finite sums).

Now

$$\sum_h E_h = \sum_h (E(x_h) - E(x_{h-1})) = I = \sum_k F_k = \sum_k (F(y_k) - F(y_{k-1}))$$

so (#) implies

$$\sum_h x_h E_h - \epsilon I \le \sum_{h,k} x_{hk} E_h F_k \le \sum_h x_h E_h + \epsilon I$$

and by Theorem 22.21

$$\left\| \sum_{h,k} x_{hk} E_h F_k - \sum_h x_h E_h \right\| \le \epsilon.$$

The sums $\sum_h x_h E_h$ converge in norm to $\int_{-\infty}^{\infty} x\, dE(x) = A$ as the "mesh" of the partition $\{(x_{h-1}, x_h]\}$ goes to 0 (25.10) so the double sums $\sum_{hk} x_{h,k} E_h F_k$ converge in norm to what we would call a double integral. Denote the double integral by $\int_{\mathbb{C}} x\, dG(z)$, $z = x + iy$. We have

$$\int_{\mathbb{C}} x\, dG(z) = \int_{-\infty}^{\infty} x\, dE(x) = A = \int_{-\infty}^{\infty} x\, dE(x) \int_{-\infty}^{\infty} dF(y).$$

Similarly, the double integral $\int_C y \, dG(z)$ exists and

$$\int_C y \, dG(z) = \int_{-\infty}^{\infty} y \, dF(y) = B = \int_{-\infty}^{\infty} y \, dF(y) \int_{-\infty}^{\infty} dE(x).$$

Thus, (*) holds with the double integral

$$\int_C z \, dG(z) = A + iB = T = \int_{-\infty}^{\infty} \int_{-\infty}^{\infty} (x + iy) \, dE(x) \, dF(y).$$

This formula represents the Spectral Theorem for normal operators. Note that we also have

$$T^* = \int_C \bar{z} \, dG(z) = A - iB = \int_{-\infty}^{\infty} \int_{-\infty}^{\infty} (x - iy) \, dE(x) \, dF(y).$$

The operational calculus for the continuous, self adjoint operators described in Chapter 26 can be extended to normal operators by defining

$$f(T) = \int_C f(t) \, dG(t)$$

for any continuous, complex valued function f vanishing outside Δ. The properties of the operational calculus described in Theorem 26.1 then carry forward.

There are also proofs of the Spectral Theorem for normal operators based on results for Banach algebras; see [TL] for an exposition.

Appendix A

Functions of Bounded Variation

For the reader who may not be familiar with the basic properties of functions of bounded variation, we record them in this appendix. Let $f : [a, b] \to \mathbb{R}$. If

$$\pi = \{a = x_0 < x_1 < \cdots < x_n = b\}$$

is a partition of $[a, b]$, the variation of f over π is

$$var\,(f : \pi) = \sum_{i=0}^{n-1} |f\,(x_{i+1}) - f\,(x_i)|,$$

and the variation of f over $[a, b]$ is

$$Var\,(f : [a, b]) = \sup var\,(f : \pi),$$

where the supremum is taken over all possible partitions, π, of $[a, b]$. If

$$Var\,(f : [a, b]) < \infty,$$

f is said to have *bounded variation*; the class of all such functions is denoted by $BV\,[a, b]$. The variation measures the amount the function oscillates in $[a, b]$.

As the example below illustrates, even a continuous function can fail to belong to $BV\,[a, b]$.

Example A.1. Let $f\,(t) = t \sin\,(1/t)$ for $0 < t \leq 1$ and $f\,(0) = 0$. Set $x_n = 1/\,(n + 1/2)\,\pi$. Then $f\,(x_n) = 1/(n + 1/2)\pi$ if n is even, and $f\,(x_n) = -1/(n + 1/2)\,\pi$ if n is odd. If π_n is the partition $\{0 < x_n < x_{n-1} < \cdots < x_1 < 1\}$, then

$$\sum_{i=1}^{n-1} |f\,(x_i) - f\,(x_{i-1})| \geq \frac{2}{\pi} \sum_{i=1}^{n-1} 1/\,(i + 1)$$

so $Var\,(f : [0, 1]) = \infty$.

Proposition A.2. *If $f \in BV[a,b]$, then f is bounded on $[a,b]$.*

Proof: Let $x \in (a,b)$. Then
$$|f(x) - f(a)| + |f(b) - f(x)| \le Var(f : [a,b])$$
so
$$2|f(x)| \le |f(a)| + |f(b)| + Var(f : [a,b])$$
and f is bounded.

We consider properties of $Var(f : I)$ as a function of the interval I. First, as a consequence of the triangle inequality, we have

Lemma A.3. *Let $f : [a,b] \to \mathbb{R}$. If π and π' are partitions of $[a,b]$ with $\pi \subset \pi'$, then $var(f : \pi) \le var(f : \pi')$.*

Proposition A.4. *Let $f : [a,b] \to \mathbb{R}$ and $a < c < b$. Then $Var(f : [a,b]) = Var(f : [a,c]) + Var(f : [c,b])$.*

Proof: Let π be a partition of $[a,b]$ and π' the partition obtained by adding the point c to π. Let π_1 and π_2 be the partitions of $[a,c]$ and $[c,b]$, respectively, induced by π'. Then by Lemma 3,

$$var(f : \pi) \le var(f : \pi') = var(f : \pi_1) + var(f : \pi_2)$$
$$\le Var(f : [a,c]) + Var(f : [c,b])$$

so
$$Var(f : [a,b]) \le Var(f : [a,c]) + Var(f : [c,b]).$$

If π_1 and π_2 partitions of $[a,c]$ and $[c,b]$, respectively, then $\pi = \pi_1 \cup \pi_2$ is a partition of $[a,b]$ so $var(f : \pi) = var(f : \pi_1) + var(f : \pi_2) \le Var(f : [a,b])$. Therefore, $Var(f : [a,c]) + Var(f : [c,b]) \le Var(f : [a,b])$.

Proposition A.5. *Let $f,g \in BV[a,b]$. Then (i) $f + g \in BV[a,b]$ with $Var(f + g : [a,b]) \le Var(f : [a,b]) + Var(g : [a,b])$, (ii) for $t \in \mathbb{R}$, $tf \in BV[a,b]$ and $Var(tf : [a,b]) = |t| Var(f : [a,b])$.*

Proof: (i) follows from the triangle inequality and (ii) is clear.

Thus, $BV[a,b]$ is a vector space under the usual operations of pointwise addition and scalar multiplication of functions.

Let $f \in BV[a,b]$. We define the total variation of f by $V_f(t) = Var(f : [a,t])$ if $a < t \le b$ and $V_f(a) = 0$.

Proposition A.6. *V_f and $V_f - f$ are increasing on $[a,b]$.*

Proof: V_f is increasing by Proposition 4.

Let $a \leq x < y \leq b$ and $g = V_f - f$. Then $g(y) = V_f(x) + Var(f : [x,y]) - f(y)$ implies

$$g(y) - g(x) = V_f(x) - f(y) + Var(f : [x,y]) - V_f(x) + f(x)$$
$$= Var(f : [x,y]) - (f(y) - f(x)) \geq 0.$$

Proposition A.7. *If $f \in BV[a,b]$ is (right, left) continuous at x, then V_f is (right, left) continuous at x.*

Proof: Let $\epsilon > 0$. Suppose f is right continuous at $x < b$. There is a partition π of $[x,b]$ such that $var(f : \pi) > Var(f : [x,b]) - \epsilon$. Since f is right continuous at x, we may add a point x_1 to π to obtain a partition $\pi' = \{x < x_1 < ... < x_n\}$ of $[x,b]$ such that $|f(x) - f(x_1)| < \epsilon$. Then

$$\epsilon + var(f : \pi') = \epsilon + |f(x) - f(x_1)| + \sum_{i=1}^{n-1} |f(x_{i+1}) - f(x_i)| < 2\epsilon + Var(f : [x_1,b])$$

so

$$Var(f : [x,b]) < var(f : \pi) + \epsilon \leq var(f : \pi') + \epsilon < 2\epsilon + Var(f : [x_1,b]).$$

Thus, $0 \leq V_f(x_1) - V_f(x) < 2\epsilon$, and since $V_f \uparrow$, $\lim_{y \to x^+} V_f(y) = V_f(x)$ and V_f is right continuous.

The other statement about left continuity is similar.

Since $f = V_f - (V_f - f)$, Propositions 6 and 7 along with Exercise 1 and Proposition 5 give a characterization of functions of bounded variation.

Theorem A.8. *Let $f : [a,b] \to \mathbb{R}$. Then $f \in BV[a,b]$ iff $f = g - h$ where $g, h \uparrow$. If f is (right, left) continuous, then g and h can be chosen to be (right, left) continuous.*

Exercises.

1. If $f \uparrow$ on $[a,b]$, show $Var(f : [a,b]) = f(b) - f(a)$.
2. Give a necessary and sufficient condition for f to satisfy $Var(f : [a,b]) = 0$.
3. Let $f, g \in BV[a,b]$. Show fg and $|f|$ belong to $BV[a,b]$.

Appendix B

The Riemann-Stieltjes Integral

In this appendix we give a brief description of the Riemann-Stieltjes integral and give statements of some of the results which will be used in the description of the dual of the space of continuous functions and in the spectral theorem. We will not include proofs of all of the results but Apostol's book ([Ap]) is an excellent reference to the integral; see also Rudin ([Ru]).

Let $f, g : [a, b] \to \mathbb{R}$. A *partition* of the interval $[a, b]$ is an ordered set $a = t_1 < t_2 < ... < t_n = b$; a *tagged partition* is a partition along with points $s_j \in [t_j, t_{j+1}]$, called *tags*. If $P = \{\{t_j\}_{j=1}^n, \{s_j\}_{j=1}^{n-1}\}$ is a tagged partition, the norm of P is defined to be $\max\{t_{j+1} - t_j : j = 1, ..., n-1\}$. The *Riemann-Stieltjes sum* of f with respect to g and the tagged partition P is

$$S(f, g, P) = \sum_{j=1}^{n-1} f(s_j)(g(t_{j+1}) - g(t_j)).$$

Definition B.1. The function f is Riemann-Stieltjes integrable with respect to g (briefly, f is g-integrable) if there exists $A \in \mathbb{R}$ such that for every $\epsilon > 0$ there exists $\delta > 0$ such that $|S(f, g, P) - A| < \epsilon$ for every tagged partition P with norm less than δ. We write $A = \int_a^b f dg = \int_a^b f(s) dg(s)$ when the integral exists.

It is easily checked that the value in Definition 1 is unique so the notation used above is justified. It also follows easily that the integral is linear in both of the variable f and g. We establish an integration by parts formulas which will be used later.

Theorem B.2. *If f is g-integrable, then g is f-integrable with*

$$\int_a^b f dg = f(b)g(b) - f(a)g(a) - \int_a^b g df.$$

Proof: Let $\epsilon > 0$ and $\delta > 0$ be as in Definition 1. Let P be a tagged partition with norm less than δ as above. Consider a Riemann-Stieltjes sum for g with respect to f and P: $S(g, f, P) = \sum_{j=1}^{n-1} g(s_j)(f(t_{j+1}) - f(t_j))$. We have

$$f(b)g(b) - f(a)g(a) = \sum_{j=1}^{n-1} f(t_{j+1})g(t_{j+1}) - \sum_{j=1}^{n-1} f(t_j)g(t_j)$$

so

$$(*) \quad f(b)g(b) - f(a)g(a) - S(g, f, P)$$
$$= \sum_{j=1}^{n-1} f(t_{j+1})(g(t_{j+1}) - g(s_j)) + \sum_{j=1}^{n-1} f(t_j)(g(s_j) - g(t_j)).$$

The two terms on the right-hand side of (*) can be combined into a single tagged partition P' by using both the points of $\{t_j\}$ and $\{s_j\}$ which has norm less than δ so

$$\left| f(b)g(b) - f(a)g(a) - S(g, f, P) - \int_a^b f \, dg \right| < \epsilon.$$

Thus, $\int_a^b g \, df$ exists and equals $f(b)g(b) - f(a)g(a) - \int_a^b f \, dg$.

We have the following sufficient condition for existence of the integral ([Ru]6.29).

Theorem B.3. *If f is continuous and g has bounded variation, then f is g-integrable and* $\left| \int_a^b f \, dg \right| \leq \|f\|_\infty \, Var(g : [a, b])$.

Later we will also need the following change of variable theorem ([Ap]7.26).

Theorem B.4. *Let f, g be continuous and $h : [a, b] \to \mathbb{R}$ have bounded variation. Then $\int_a^b f(s) d(\int_a^s g(t) dh(t)) = \int_a^b f(s)g(s) dh(s)$.*

Appendix C

The Dual of $C[a, b]$

In this appendix we describe the dual of the space $C[a, b]$ of continuous real valued functions defined on the interval $[a, b]$ equipped with the sup-norm. The description of the dual uses functions of bounded variation and the Riemann-Stieltjes integral which are discussed in Appendices A and B.

Theorem C.1. *Let $G \in C[a, b]'$. Then there exists a function $g : [a, b] \to \mathbb{R}$ of bounded variation such that $G(f) = \int_a^b f dg$ for every $f \in C[a, b]$. Moreover, $\|G\| = Var(g : [a, b])$.*

Proof: The space $C[a, b]$ is a closed subspace of the Banach space $B[a, b]$ (Example 1.6) so by Theorem 7.1 the functional G can be extended to a continuous linear functional on $B[a, b]$, still denoted by G, with the same norm. Define $g : [a, b] \to \mathbb{R}$ by $g(t) = G(\chi_{[a,t]})$ for $a < t \leq b$ and $g(a) = 0$, where $\chi_{[a,t]}$ is the characteristic function of $[a, t]$. We first show that g has bounded variation with $Var(g : [a, b]) \leq \|G\|$. Let $a = t_1 < ... < t_n = b$ be a partition of $[a, b]$. Set $s_j = sign(g(t_{j+1}) - g(t_j))$ and $f = s_1 \chi_{[t_1, t_2]} + \sum_{j=2}^{n-1} s_j \chi_{(t_j, t_{j+1}]}$ so $f \in B[a, b]$ and $\|f\|_\infty = 1$. Then

$$G(f) = s_1 G(\chi_{[a,t_2]} - \chi_{[a,t_1]}) + \sum_{j=2}^{n-1} s_j G(\chi_{[a,t_{j+1}]} - \chi_{[a,t_j]})$$

$$= \sum_{j=1}^{n-1} s_j(g(t_{j+1}) - g(t_j)) = \sum_{j=1}^{n-1} |g(t_{j+1}) - g(t_j)| \leq \|G\|$$

which implies $Var(g : [a, b]) \leq \|G\|$.

Let $f \in C[a, b]$. We show $G(f) = \int_a^b f dg$; note there is no problem with the existence of the integral since f is continuous and g is of bounded variation [see Appendix B]. Let P be the partition above and define $f_P = f(t_1)\chi_{[a,t_1]} + \sum_{j=2}^{n-1} f(t_j)\chi_{(t_j,t_{j+1}]}$. If $\delta = \max\{|t_{j+1} - t_j| : j = 1, ..., n-1\}$

is the norm of the partition, note that $f_P \to f$ in $\|\cdot\|_\infty$ as $\delta \to 0$ by the uniform continuity of f. Now $G(f_P) = \sum_{j=1}^{n-1} f(t_j)(g(t_{j+1}) - g(t_j))$ is a Riemann-Stieltjes sum with respect to the partition P and the points $\{t_j\}$ so $G(f_P)$ converges to $\int_a^b f dg$ as $\delta \to 0$ and $G(f_P)$ converges to $G(f)$ by the continuity of G; this gives the representation. Since the inequality $|G(f)| = \left|\int_a^b f dg\right| \leq \|f\|_\infty Var(g : [a,b])$ holds for any Riemann-Stieltjes integral (Appendix B), $\|G\| \leq Var(g : [a,b])$.

Theorem 1 gives a linear, norm preserving map from the dual of $C[a,b]$ into $BV[a,b]$ (see Appendix A and Example 1.10); however, the map is not one-one since if g is the function in the statement of Theorem 1, then the function $g + c$ for any constant c also satisfies the representation. In order to obtain a one-one map we need to single out a special subspace of $BV[a,b]$. To accomplish this we begin by defining an equivalence relation on $BV[a,b]$.

Two functions g and h in $BV[a,b]$ are equivalent, $g \sim h$, if $\int_a^b f dg = \int_a^b f dh$ for every $f \in C[a,b]$. It is easily seen that this defines an equivalence relation on $BV[a,b]$.

Lemma C.2. $g \sim 0$ iff $g(a) = g(b)$ and $g(t^+) = g(t^-) = g(a)$ for $a < t < b$.

Proof: Suppose $g \sim 0$. First note $\int_a^b dg = g(b) - g(a) = 0$. Next, note that if $v \in BV[a,b]$, then $\int_t^{t+h} dv/h \to v(t^+)$ as $h \to 0^+$ for $a \leq t < b$ and $\int_{t-h}^t dv/h \to v(t^-)$ as $h \to 0^+$ for $a < t \leq b$. For $a \leq t < t + h < b$, define $f(s) = 1$ if $a \leq s \leq t$, $f(s) = 1 - (s-t)/h$ if $t \leq s \leq t + h$ and $f(s) = 0$ if $t + h \leq s \leq b$. Then $f \in C[a,b]$ and integration by parts (see Appendix B) gives

$$0 = \int_a^b f dg = g(t) - g(a) + \int_t^{t+h} f(s) dg(s)$$

$$= g(t) - g(a) - g(t) + \int_t^{t+h} g(s) ds/h$$

so $g(t^+) = g(a)$ by the observation above. Similarly, $g(t^-) = g(b)$ if $a < t \leq b$.

If g satisfies the conditions of the lemma, then g is equivalent to the constant function $g(a)$ since both of these functions are equal at all interior points of $[a,b]$ where g is continuous and then $\int_a^b f dg = \int_a^b f dg(a) = 0$. [The Riemann-Stieltjes sums for these functions can be defined by using only the points of continuity of g.]

We now use the equivalence relation defined above to single out a sub-space of $BV[a,b]$ which will be isometrically isomorphic to the dual of $C[a,b]$. The function $g \in BV[a,b]$ is *normalized* if $g(a) = 0$ and g is right continuous on (a,b). From Lemma 2 if g and h are normalized and $g \sim h$, then $g = h$. If $g \in BV[a,b]$ and if we define $g\hat{\ }$ by $g\hat{\ }(a) = 0$, $g\hat{\ }(b) = g(b) - g(a)$ and $g\hat{\ }(t) = g(t^+) - g(a)$ for $a < t < b$, then $g \sim g\hat{\ }$. Thus, each equivalence class of $\sim$ contains exactly one normalized function of bounded variation. We also claim that $Var(g : [a,b]) \geq Var(g\hat{\ } : [a,b])$. For suppose $a = t_1 < ... < t_n = b$ is a partition of $[a,b]$. For $\epsilon > 0$ small, we can choose points s_j, $j = 1,...,n-1$, to the right of t_j with $s_j < t_{j+1}$ and $\left|g(t_j^+) - g(s_j)\right| < \epsilon/2n$. If we set $s_0 = a$ and $s_n = b$, we have $\sum_{j=1}^{n-1} \left|g\hat{\ }(t_{j+1}) - g\hat{\ }(t_j)\right| \leq \sum_{j=1}^{n-1} |g(s_{j+1}) - g(s_j)| + \epsilon \leq Var(g : [a,b]) + \epsilon$ so $Var(g\hat{\ } : [a,b]) \leq Var(g : [a,b])$.

We denote the subspace of normalized functions of bounded variation by $NBV[a,b]$ and now show that the dual of $C[a,b]$ is isometrically isomorphic to $NBV[a,b]$. Note that the norm on $NBV[a,b]$ is just $\|g\| = Var(g; [a,b])$ since $g(a) = 0$.

Theorem C.3. *The linear mapping $\Psi : NBV[a,b] \to C[a,b]'$ defined by $\Psi g(f) = \int_a^b f \, dg$, $f \in C[a,b]$, is an isometry onto $C[a,b]'$.*

Proof: If $g \in NBV[a,b]$, then $|\Psi g(f)| = \left|\int_a^b f \, dg\right| \leq \|f\|_\infty Var(g : [a,b])$ so Ψ is a linear continuous map from $NBV[a,b]$ into $C[a,b]'$ with $\|\Psi g\| \leq Var(g : [a,b])$. We claim that Ψ is onto $C[a,b]'$. If $G \in C[a,b]'$, then by Theorem 1 there exists a function of bounded variation g such that $G(f) = \int_a^b f \, dg$ for every $f \in C[a,b]$ and $\|G\| = Var(g : [a,b])$. We then have $G(f) = \int_a^b f \, dg = \int_a^b f \, dg\hat{\ } = \Psi g\hat{\ }(f)$ so $\Psi g\hat{\ } = G$ and $\|\Psi g\hat{\ }\| = \|G\| \leq Var(g\hat{\ } : [a,b]) \leq Var(g : [a,b]) = \|G\|$. Thus, Ψ is a linear isometry onto $C[a,b]'$.

For later use we will need the following density result.

Theorem C.4. *Let $g \in NBV[a,b]$. If $\int_a^b t^n \, dg(t) = 0$ for $n = 0, 1, ...,$ then $g = 0$.*

Proof: Let G be the continuous linear functional on $C[a,b]$ induced by g. Then $Gp = 0$ for every polynomial p. Thus, $Gf = 0$ for every continuous function f by the Weierstrass Approximation Theorem and the continuity of G.

Exercises.

1. Define $G : C[0,1] \to \mathbb{R}$ by $G(f) = f(0)$. Show $G \in C[0,1]'$ and find the function g in Theorem 3.

2. Let $a \in C[0,1]$ and define $G : C[0,1] \to \mathbb{R}$ by $G(f) = \int_0^1 af$. Show $G \in C[0,1]'$ and find the function g in Theorem 3.

Appendix D

The Baire Category Theorem

The Baire Category Theorem will be used several times in the text. For the convenience of the reader unfamiliar with this result, we give a statement and proof in this appendix. References to further applications of the Baire Category Theorem are given at the end of the appendix.

Let (S, d) be a metric space. A subset $E \subset S$ is *nowhere dense* if the interior of $\overline{E}$, the closure of E, is empty. For example, the Cantor set is nowhere dense in $[0, 1]$ [Exercise 2]. A subset $E \subset S$ is *first category* in S if E is a countable union of nowhere dense sets. For example, $\mathbb{Q}$ is first category in $\mathbb{R}$. A subset $E \subset S$ is *second category* in S if E is not first category in S. Baire's Theorem asserts that every complete metric space is second category in itself.

Proposition D.1. *Let $\{F_k\}$ be a sequence of closed sets contained in the complete metric space (S, d) such that $F_k \supset F_{k+1}$ and $d_k = diameter F_k \to 0$. Then $\cap_{k=1}^{\infty} F_k$ is a singleton.*

Proof : It suffices to show $\cap_{k=1}^{\infty} F_k \neq \emptyset$. Pick $x_k \in F_k$. Then $d(x_k, x_j) \leq d_k$ for $j \geq k$ so $\{x_j\}$ is Cauchy and, therefore, convergent to some $x \in S$. Clearly $x \in \cap_{k=1}^{\infty} F_k$.

Theorem D.2. *(Baire Category Theorem) A complete metric space is second category in itself.*

Proof : Let A_k be nowhere dense in S for every $k \in N$ and set $E = \cup_{k=1}^{\infty} A_k$. We show $S \backslash E \neq \emptyset$. Let $x_o \in S$. Since A_1 is nowhere dense, there is a closed ball B_1 of radius less than $1/2$ inside the closed ball $B_0 = \{x : d(x, x_0) \leq 1\}$ such that $B_1 \cap A_1 = \emptyset$ (Exercise 1). Since A_2 is nowhere dense, there is a closed ball B_2 of radius less than $1/2^2$ inside B_1 such that $B_2 \cap A_2 = \emptyset$. Continuing this construction gives a decreasing sequence of

173

closed balls $\{B_k\}$ of radius less than $1/2^k$ such that $B_k \cap A_k = \emptyset$ for all k. By Proposition 1, $\cap_{k=1}^{\infty} B_k = \{x\}$. Clearly $x \in S \backslash E$.

We give a corollary of the Baire Category Theorem which is particularly useful in applications.

Corollary D.3. *Let* (S, d) *be a complete metric space. If* $S = \cup_{k=1}^{\infty} A_k$, *then some* $\overline{A_k}$ *must have a non-void interior.*

Despite its esoteric appearance the Baire Category Theorem has a surprising number of applications to various areas of analysis. For example, Banach used the theorem to show the existence of a continuous, nowhere differentiable function. For this and other interesting examples, see [DeS] and [Boa].

Exercises

1. Show that E is nowhere dense if and only if every sphere S contains a sphere S' such that $S' \cap E = \emptyset$.

2. Show that Cantor set is nowhere dense in $[0, 1.]$

3. Let $f_k : \mathbb{R} \to \mathbb{R}$ be continuous, nonnegative and such that $\sum_{k=1}^{\infty} f_k(t)$ converges for every $t \in \mathbb{R}$. Show that there is an interval in $\mathbb{R}$ where the convergence is uniform.

4. Show that $\mathbb{R}^2$ is not a countable union of lines.

5. Show that if S is a complete metric space without isolated points, then S is uncountable.

References

[AA] Y.A. Abrowovich and C.D. Aliprantis, An Invitation to Operator Theory, Amer. Math. Soc., Providence, 2002.

[Ah] L. Ahlfors, Complex Analysis, McGraw-Hill, N.Y., 1953.

[Ap] T. Apostol, Mathematical Analysis, Addison-Wesley, Reading, 1974.

[BN] G. Bachman and L. Narici, Functional Analysis, Academic Press, N.Y., 1966.

[B] S. Banach, Oeuvres II, Acad. Pol. Sci, Warsaw, 1979.

[Ba] R. Bartle, The Elements of Integration, Wiley, N.Y., 1966.

[BeK] G. Bennett and N. Kalton, Consistency Theorems for Almost Convergence, Trans. Amer. Math. Soc., 198(1974), 23-43.

[BK] G. Birkoff and E. Kreyszig, The Establishment of Functional Analysis, Historia Math., 11(1984), 258-321.

[Boa] R P. Boas, A Primer of Real Functions, Math. Assoc. Amer., Providence, 1960.

[CB] R. Churchill and J.W. Brown, Complex Variables and Applications, McGraw-Hill, N.Y., 1984.

[Co] J. Conway, A Course in Functional Analysis, Springer-Verlag, N.Y., 1985.

[DM] L. Debnath and P. Mikusinski, Introduction to Hilbert Space and Applications, Adademic Press, San Diego, 1990.

[DeS] J. DePree and C. Swartz, Introduction to Real Analysis, Wiley, N.Y., 1988.

[Di] J. Dieudounne, History of Functional Analysis, North-Holland, Amsterdam, 1981.

[DuS] N. Dunford and J. Schwartz, Linear Operators I, Wiley, N.Y.,

1958.

[En] P. Enflo, A counterexample to the approximation problem in Banach spaces, Acta Math., 130(1973), 309-317.

[En2] P. Enflo, On the Invariant Subspace Problem in Banach Spaces, Acta Math., 158(1987), 213-313.

[Go] S. Golberg, Unbounded Linear Operators, McGraw-Hill, N.Y., 1966.

[Gr] D.H. Griffel, Applied Functional Analysis, Wiley, N.Y., 1981.

[HS] E. Hewitt and K. Stromberg, Real and Abstract Analysis, Springer-Verlag, N.Y., 1965.

[Ho] H. Hochstadt, Eduard Helly, Father of the Hahn-Banach Theorem, Math. Intell., 2(1979), 123-125.

[J] R.C. James, A non-reflexive Banach space isometric with its second conjugate space, Proc. Nat. Acad. Sci. U.S.A., 37(1951), 174-177.

[J2] R.C. James, Bases in Banach Spaces, Amer. Math. Monthly, 89(1982), 625-641.

[LT] J. Lindenstrauss and L. Tzafriri, Classical Banach Spaces I and II, Springer-Verlag, Berlin, 1977.

[Lu] D. Luenberger, Optimization by Vector Space Methods, Wiley, N.Y., 1969.

[Ma] I. Maddox, Elements of Functional Analysis, Cambridge Univ. Press, 1970.

[Me] R. Megginson, An Introduction to Banach Space Theory, Springer-Verlag, N.Y., 1998.

[Mon] A. F. Monna, Functional Analysis in Historical Perspective, Wiley, N.Y., 1973.

[Mo] T.J. Morrison, Functional Analysis, Wiley, N.Y., 2001.

[Na] I. P. Natanson, Constructive Function Theory (translation), Atomic Energy Commision, 1961.

[Pr] J. D. Pryce, Basic Methods of Linear Functional Analysis, Hutchinson Univ. Library, London, 1973.

[RN] F.Riesz and B. Sz-Nagy, Functional Analysis, Ungar, N.Y. 1955.

[Ru] W. Rudin, Principles of Mathematical Analysis, McGraw-Hill, N.Y., 1964.

[Ru2] W. Rudin, Functional Analysis, McGraw-Hill, N.Y., 1991.

[Sch] M. Schecter, Principles of Functional Analysis, Amer. Math. Soc., Providence, 2002.

[SS] R. Siegmund-Schultze, The Origins of Functional Analysis, A History of Analysis, Amer. Math. Soc., 2003, 385-404.

[Sm] F. Smithies, The Shaping of Functional Analysis, Bull. London Math. Soc., 29(1997), 129-138.

[Ste] L. Steen, Highlights in the History of Spectral Theory, Amer. Math. Monthly, 80(1973), 359-381.

[St] R. Strichartz, How to Make Wavelets, Amer. Math. Monthly, 100(1993), 539-556.

[Sw1] C. Swartz, Measure, Integration and Function Spaces, World Sci. Publ., Singapore, 1994.

[Sw2] C. Swartz, The Evolution of the Uniform Boundedness Principle, Math. Chron., (1990), 1-18.

[TL] A. Taylor and D. Lay, Introduction to Functional Analysis, Wiley, N.Y., 1980.

[Wa] D. Walnut, Wavelet Analysis, Birkhauser, Boston, 2002.

[Wh] R. Whitley, Projecting m onto c_o, Amer. Math. Monthly, 73 (1966), 285-286.

[Y] K.Yosida, Functional Analysis, Springer-Verlag, Berlin, 1966.

Index

$B(S)$, 2
$BV[a, b]$, 3
$C(S)$, 2
$C\sigma(T)$, 107
$C^1[a, b]$, 2
$L^p(E)$, 6
$M(T)$, 116
M^0, 86
$NBV[a, b]$, 171
N_0, 86
$P\sigma(T)$, 107
$R\sigma(T)$, 107
$\mathbb{R}^n$, 2
$\rho(T)$, 103
$\sigma(T)$, 103
bs, 3
c, 2
$c_{00}(c_c)$, 2
c_0, 2
e^j, 5
l^∞, 2
l^p, 3
$m(T)$, 116
m_0, 2
$r(T)$, 106
s, 2
$\mathcal{R}[a, b]$, 7
T^*, 89
$Var\,(f : [a, b])$, 163
$J_X = J$, 49
$L(X, Y)$, 9
$\mathcal{D}T$, 103

X', 9
$\Bbbk$, 1
$BV\,[a, b]$, 163
$C^n[a, b]$, 7
$\mathcal{N}T$, 85
$M^\perp$, 33
$x \perp y$, 32
$\mathcal{R}T$, 85
$R_\lambda(T) = R_\lambda$, 104
$V_f(t)$, 164
T', 83

absolutely convergent, 6
adjoint, 89
AK-space, 81
annihilator, 22, 86
approximate eigenvalue, 108
approximate spectrum, 108
approximation property, 94
automatic continuity, 55

Baire Category Theorem, 173
Banach limit, 43
Banach space, 4
Banach-Steinhaus, 54
Bessel's Inequality, 34
bidual, 48
BK-space, 81
Bohnenblust/Sobczyk, 42
bounded, 7
bounded operator, 9
bounded series, 3

bounded variation, 163

canonical imbedding, 49
Cauchy-Schwarz, 4
Cesaro matrix, 17
Closed Graph Theorem, 70
compact operator, 91
complementary, 73
complete, 35
completely continuous, 95
completion, 50
conservative, 56
continuous spectrum, 107
convergent series, 3
convex, 31
coordinate functionals, 77
countably subadditive, 69

Dirichlet kernel, 56
dual norm, 10
dual space, 9
Dvoretsky/Rogers, 27

eigenvalue, 107
eigenvector, 107
equivalent, 25
extended limit, 43

first category, 173
Fourier coefficients, 34
Fredholm Alternative, 101
Fredholm equation, 14, 97

generalized Schwarz inequality, 137
Gram-Schmidt, 38
graph, 70
Green's function, 131

Haar system, 38
Hahn-Banach, 41
Hahn-Schur, 63
Hellinger-Toeplitz, 115
Hilbert space, 30
Hilbert/Schmidt, 123
Holder, 3

inner product, 29
inner product space, 29
integral operator, 11
integration by parts, 167
invariant, 113
isometry, 15
isomorphic, 25

K-space, 81
kernel, 11

Lagrange interpolation polynomial,
    58
Legendre polynomials, 38

Mercer, 124
Minkowski, 3
moments, 44
monotone, 78

Neumann series, 12
norm, 1
normal operator, 135
normalized, 171
nowhere dense, 173

Open Mapping Theorem, 71
operational calculus, 151
operator semi-norm, 10
orthogonal, 32
orthogonal complement, 32
orthogonal projection, 141
orthonormal, 32
orthonormal basis, 35
orthonormal dimension, 37

parallelogram law, 30
Parseval's Equality, 36
partition, 167
point spectrum, 107
polar decomposition, 154
positive operator, 117, 134
precompact operator, 91
predual, 90
projection, 73

quotient space, 21

reciprocal kernel, 15
reflexive, 49
regular, 56
residual spectrum, 107
resolution of the identity, 149
resolvent equation, 105
resolvent operator, 104
resolvent set, 103
Riemann-Stieltjes integral, 167
Riemann-Stieltjes sum, 167
Riesz, 26
Riesz Representation Theorem, 33

Schauder, 94
Schauder basis, 77
Schwarz Inequality, 29
self adjoint, 133
semi-norm, 1
separable, 7
sesquilinear, 143
Silvermann-Toeplitz, 56
Spectral Mapping Theorem, 106, 157
spectral radius, 106
spectral theorem, 145

spectrum, 103
square root, 153
Steinhaus, 65
Sturm-Liouville operator, 116, 130
sublinear, 41
subseries convergent, 27
summation operator, 11
sup-norm, 2
symmetric, 115

tagged partition, 167
tags, 167
transpose, 83

Uniform Boundedness Principle, 53
unitary, 139

Volterra, 93
Volterra operator, 11

weak convergence, 61
weak* convergent, 65
weakly Cauchy, 62
weakly sequentially complete, 62

Zabreiko, 69